Singularities of Functions, W ....,
Caustics and Multidimensional Integrals

To Bob

Happy Christmas and
best wishes for
New Year - 2023 !

Jamie

Singularities of Functions, Wave Fronts, Caustics and Multidimensional Integrals

V.I.Arnold, A.N.Varchenko, A.B.Givental and A.G.Khovanskii

C S P

© 2012 Cambridge Scientific Publishers

Reviews in Mathematics and Mathematical Physics
Volume 14, Part 2

First published in Soviet Scientific Reviews: Section C © 1984 OPA (Overseas Publishers Association) N.V. Published by license under the Harwood Academic Publishers imprint.

Printed in UK

ISBN 978-1-904868-98-9 Paperback

Acknowledgements/Credits
Photographs from MCCME/IUM

Cambridge Scientific Publishers Ltd
45 Margett Street, Cottenham, Cambridge CB24 8QY UK
www.cambridgescientificpublishers.com

Introduction to the Series

In recent years, many young excellent mathematicians both in Russia and abroad made an excellent name for themselves in the international mathematical community. Reviews in Mathematics and Mathematical Physics will publish the most outstanding and recent results. Not only new research, but also a detailed background of a problem will be presented to give an insight into a particular direction. This would enable a better grasp of the state of the art in a particular field and an easier positioning of the results discussed.

To maintain high scientific standards, all works published in Reviews in Mathematics and Mathematical Physics are peer reviewed by two or three internationally recognized mathematicians. Many of the works presented by Russian mathematicians have been discussed at the seminars at the Faculty of Mechanics and Mathematics, Moscow State University and Steklov Mathematical Institute, Russian Academy of Sciences.

Reviews in Mathematics and Mathematical Physics also plan publications of review papers by oustanding mathematicians to ensure a broader view of the development of modern mathematics and mathematical physics.

A.T.Fomenko
Editor

Vladimir Igorevich Arnold (1937-2010)

Preface

This edition is a tribute to Vladimir Igorevich Arnold (1937-2010) in recognition of his achievements and significant contribution to mathematics. He was a world-class mathematician and scientist, a dedicated teacher and lecturer and was the author of several textbooks including Mathematical Methods of Classical Mechanics.In 1990, he became an Academician of Russian Academy of Sciences and he was one of the founders of the Independent University of Moscow which was established in 1991, and continues to be one of the leading mathematical centres in Russia.

Vladimir Igorevich Arnold was born in 1937 in Odessa, now part of Ukraine. He was an outstanding student of Moscow State University; he graduated in 1959 and in 1961, under the supervision of Professor A.N.Kolmogorov, he defended his PhD thesis entitled: "On the representation of continuous functions of 3 variables by the superpositions of continuous functions of 2 variables". His thesis solved 13th Hilbert Problem, thereby establishing his reputation and the beginning of his distinguished career. V.I.Arnold was appointed Professor of Mathematics at Moscow State University from 1965 to 1986. He then joined the Steklov Mathematical Institute in Moscow and in 1993 he was appointed Professor at University of Paris-Dauphine, and he worked at these two institutes, dividing his time between Paris and Moscow, spring and summer in Paris and fall and winter in Moscow.

The scope of Professor V.I. Arnold's mathematical interest was wide ranging and he made important contributions to modern mathematics in several areas including dynamical systems theory, classical and celestial mechanics, mathematical physics, variational calculus, real and complex algebraic geometry, differential geometry, topology. He was one of the founders of singularity theory/catastrophe theory (Rene Thom was also a founder) which predicts how under certain circumstances slow smooth changes in a system can lead to a sudden and major change. The theory has applications in physics, chemistry and biology. Professor Andrei Kolmogorov published the first work in 1954; and later in the 1960s together with Professor Vladimir Arnold and Professor Jurgen Moser, made fundamental discoveries in the study of stability in dynamic systems using the motion of the planets around the sun as an example. The introduction of a third body (the moon) gives rise to the famous 3-body problem in which the motion of each body seems impossible to predict. The impressive work resulting in the well known Kolmogorov-Arnold-Moser (KAM) theory laid down proofs for certain circumstances and established a framework for future studies.

Professor V.I.Arnold was an established author and editor and published extensively including some 700 scientific papers, monographs and text books. Arnold's Mathematical Methods of Classical Mechanics and Ordinary Differential Equations are well known textbooks central to the education of thousands of modern scientists and their introduction to solving ordinary differential equations. He was an inspired teacher and lecturer and could present difficult theories in a clear and simple way. V.I.Arnold received many awards during his distinguished career including: the Lenin Prize (1965 with Andrei Kolmogorov), the Harvey Prize (1994), the Dannie Heineman Prize for Mathematical Physics in 2001, the Wolf Prize in Mathematics in 2001, the State Prize of the Russian Federation in 2007 and the Shaw Prize in Mathematical Sciences in 2008.

Vladimir Igorevich Arnold was one of the greatest and most influential mathematicians of the twentieth century and the significance of his contribution to mathematics is evident and his memory will endure.

Singularities of Functions, Wave Fronts, Caustics and Multidimensional Integrals – V.I.Arnold, A.N.Varchenko, A.B.Givental and A.G.Khovanskii (Moscow State University). The first edition of this paper was published in 1984 in Soviet Scientific Reviews: Mathematical Physics Reviews: Volume 4: a compiled volume edited by Professor S.P.Novikov. His foreword described the paper as "an introduction to some difficult contemporary fields of study in mathematics known under the rubric of Catastrophe Theory, which encompasses the theory of typical singularities of functions and mappings".

Vladimir Igorovich Arnold
Selected Bibliography

V. I. Arnold, Mathematical Methods of Classical Mechanics, Springer-Verlag (1989), ISBN 0-387-96890-3.
V. I. Arnold, Geometrical Methods In The Theory Of Ordinary Differential Equations, Springer-Verlag (1988), ISBN 0-387-96649-8.
V. I. Arnold, Ordinary Differential Equations, The MIT Press (1978), ISBN 0-262-51018-9.
V. I. Arnold, A. Avez, Ergodic Problems of Classical Mechanics, Addison-Wesley (1989), ISBN 0-201-09406-1.
V. I. Arnold, Teoriya Katastrof (Catastrophe Theory, in Russian), 4th ed. Moscow, Editorial-URSS (2004), ISBN 5-354-00674-0.
V. I. Arnold, "Tsepniye Drobi" (Continued Fractions, in Russian), Moscow (2001).

V. I. Arnold, Yesterday and Long Ago, Springer (2007), ISBN 978-3-540-28734-6.

Arnold, V. I.; V. S. Afraimovich (1999). Bifurcation Theory And Catastrophe Theory. Springer. ISBN 3540653791.

Arnold, V. I.: On the teaching of mathematics. (Russian) Uspekhi Mat. Nauk 53 (1998), no. 1(319), 229--234; translation in Russian Math. Surveys 53 (1998), no. 1, 229--236.

V. I. Arnold, M. Atiyah, P. Lax, B. Mazar: Mathematics: frontiers and perspectives. International Mathematical Union (2000) ISBN 0-8218-2070-2

Vladimir Igorovich Arnold: Selected Publications

1. On the representability of functions of two variables in the form $X\chi(\phi(\Phi)+\psi(\Psi))$. Uspehy Math. Nauk 1957, 12:2, 119-121.

2. On the functions of three variables. Doklady AN USSR, 1957, 114:4, 679-681 English transl. Amer Math Soc. Trans (2) 28 (1963) 51-54.

3. On the representation of continuous functions of three variables by the superpositions of continuous functions of two variables. Matem. Sbornik, 1959, 48:1, 3-74 and 1962, 56:3, 392. English transl. Amer Math Soc. Trans (2) 28 (1963) 61-147.

4. A criterion of the nomografibility on the rectangular Cartesian abacus. Uspehy Math. Nauk, 1961, 16:4, 133-135.

5. Small denominators. I: On the maps of a circle onto itself. Izvestija Ac. Sci. USSR, Ser. Math., 1961, 25:1, 21-86 and 1964, 28:2, 479-480.

6. On the stability of the equilibrium of a Hamiltonian system of ordinary differential equations in a generic elliptic case. Doklady AN USSR, 1961, 137:2, 255-257.

7. On the birth of a conditional-periodic motion from a family of periodic motions. Doklady, 1961, 138:1, 13-15.

8. Some remarks on the flows of linear elements and frames. Doklady, 1961, 138:2, 255-257.

9.   Notes on the rotation numbers. Siberian Math. Jour., 1961, 2:6, 807-813.

10. On the behavior of adiabatic invariants under a slow periodic change of the Hamiltonian function. Doklady, 1962, 142:4, 758-761.

11. On small perturbations of automorphisms of tori (with Ya.G. Sinai), Doklady, 1962, 144:4, 695-698.

12. On the classical perturbation theory and stability theory of planetary systems. Doklady, 1962, 145:3, 487-490.

13. A proof of the A.N. Kolmogorov's theorem on the conservation of conditional-periodic motions in a small change of the Hamiltonian function. Uspehy Math. Nauk, 1963, 18:5, 13-40.

14. Small denominators and problems on the stability of motions in the classical and celestial mechanics. Uspehy Math. Nauk, 1963, 18:6, 91-192.

15. On one theorem of Liouville, concerning integrable problems of dynamics. Siberian Math. J., 1963, 4:2, 471-474.

16. Homogeneous distribution of points on a sphere and some ergodic properties of linear ordinary differential equations in the complex domain (with A.L. Krylov). Doklady, 1963, 148:1, 9-12.

17. On the nonstability of dynamical systems with many degrees of freedom. Doklady, 1964, 156:1, 9-12.

18. Conditions of the applicability and an estimate of the mistake of the averaging method for systems, which goes through the resonances during the evolution process. Doklady, 1965, 161:1, 9-12.

19. On the conditions of the nonlinear stability of flat stationary curvilinear flows of the ideal fluid. Doklady, 1965, 162:5, 975-978.

20. A Variational principle for three-dimensional stationary flows of the ideal fluid. Applied Math. and Mechan., 1965, 29:5, 846-851.

21. On the topology of three-dimensional stationary flows of the ideal fluid. Applied Math. and Mech., 1966, 30:1, 183-185.

22. Sur la courbure de Riemann des groupes de diffeomorphismes, C.R.Ac.Sci. Paris, 1965, v.260, 5668-5671.

23. Sur la topologie des ecoulements stationnaires des fluides parfaits, C.R.Ac.Sci. Paris, 1965, v.261, 17-20.

24. Sur une propriete topologique des applications globalement canoniques de la mecanique classique, C.R.Ac. Sci.Paris, 1965, 261, 3719-3722.

25. Sur la geometrie differentielle des groupes de Lie de dimension infinie et ses applications a l'hydrodynamique des fluides parfaits. Ann. Inst. Fourier, 1966, 16:1, 319-361.

26. Sur un principe variationel pour les ecoulements stationaires. J. de Mecanique, Paris, 1966, 5:1, 29-43.

27. Problemes ergodiques de la mecanique classique (with A. Avez). Gauthier-Villars, 1967, a book of 243 pages.

28. On one a priori estimate in the theory of the hydrodynamical stability. Izvestija VUZov, Ser.Mat., 1966, No.5, 3-5.

29. Stability problem and ergodic properties of classical dynamical systems. Proc. Intern. Congr. of Math. (Moscow, 1966); Trans. Congr. Intern. of Mathematicians (Moscow, 1966). Mir Publishers, 1968, 387-392.

30. On a characteristic class entering in the quantization conditions. Funct. Anal. and its Appl. (FAA), 1967, 1:1, 1-14.

31. A note on the Weierstrass preparation theorem. FAA, 1967, 1:3, 1-8.

32. Singularities of smooth mappings. Uspehy Math. Nauk, 1968, 23:1,3-44.

33. On braids of algebraic functions and cohomologies of swallowtails. Uspehy Math. Nauk, 1968, 23:4, 247-248.

34. A remark on the ramification of hyperelliptic integrals as functions of parameters. FAA 1968, 2:3, 1-3.

35. Notes on the singularities of finite codimension in the complex dynamical systems. FAA, 1969, 3:1, 1-6.

36. The cohomology ring of the group of colored braids. Mat.Zametki (Math. Notes), 1969, 5:2, 227-231.

37. On some topological invariants of algebraic functions, I. Trans. Moscow Math. Society, 1970, V.21, 27-46.

38. On one-dimensional cohomology of the Lie algebra of nondivergent vector fields and rotation numbers of dynamical systems. FAA, 1969, 3:4, 77-78.

39. Hamiltonial character of the Euler equations of dynamics of solids and of ideal fluid. Uspehy Math. Nauk, 1969, 24:6, 225-226.

40. On cohomology classes of algebraic functions, stable under Tschirnhausen transforms. FAA, 1970, 4:1, 84-85.

41. Topological invariants of algebraic functions, II. FAA, 1970, 4:2, 1-9.

42. On local problems of Analysis. Vestnik Mosc. Univ., Ser. Math., 1970, No.2, 52-56.

43. Algebraic nonsolvability of the problem of Ljapunov stability and of the problem of the topological classification of singular points of analytic systems of differential equations. Uspehy Math. Nauk, 1970, 25:2, 265-266.

44. (The same title as for No.43) FAA, 1970, 4:3, 1-9.

45. On matrices depending on parameters. Rus. Math. Surv., 1971, 26:2, 101-114.

46. On the dispositions of ovals of real plane algebraic curves, involutions of four-dimensional smooth manifolds and arithmetics of integer quadratic forms. FAA 1971, 5:3, 1-9.

47. Ordinary differential equations. Moscow, Nauka, 1971, 1-240.

48. Notes on the behavior of flows of the three-dimensional ideal fluid under a small perturbation of the initial velocity field. Appl. Math. Mech. 1972, 36:2, 255-262.

49. A comment to "Sur un theoreme de la geometrie". In: Izbrannye trudy A.Puankare (H. Poincare, Selected Works), M., Nauka, 1972, vol.2, 987-989.

50. Integrals of rapidly oscillating functions and singularities of projections of Lagrange manifolds. FAA 1972, 6:3, 61-62.

51. Normal forms of functions near degenerate critical points, Weyl groups A, D, E and Lagrange singularities. FAA 1972, 6:4, 3-25.

52. Lectures on bifurcations and versal families. Rus. Math. Surv. 1972, 27:5, 119-182.

53. Modes and quasimodes. FAA 1972, 6:2, 12-20.

54. Classification of unimodal critical points of functions. FAA 1973, 7:3, 75-76.

55. Notes on the stationary phase method and Coxeter numbers. Uspehy Math. Nauk, 28:5, 1973, 17-44.

56. Normal forms of functions in a neighborhood of degenerate critical points. Rus. Math. Surv. 1974, 29:2, 11-49.

57. Topology of real algebraic curves (works of I.G.Petrovsky and their development). Uspehy Math. Nauk, 1974, 28:5, 260-262.

58. Critical points of functions and classification of caustics. Uspehy Math. Nauk, 1974, 29:2, 243-244.

59. Mathematical methods of classical mechanics. Moscow, Nauka, 1974, 432p.

60. Asymptotical Hopf invariant and its applications. Trans. of All-Union School on differential equations. Erevan, 1974, 229-256 (Engl. translation: Selecta Math. Sov., 1986, 5:4, 327-346).

61. Critical point of smooth functions. Vancouver Intern. Congr. of Math., 1974, vol.1, 19-39.

62. Classification of bimodal critical points of functions. FAA 1975, 9:1, 49-50.

63. Critical points of smooth functions and their normal forms. Rus. Math. Surv., 1975, 30:5, 3-65.

64. Local normal forms of functions. Invent. Math. 1976, 35:1, 87-109.

65. A spectral sequence for the reduction of functions to normal forms. FAA 1975, 9:3, 81-82.

66. Spectral sequences for reduction of functions to normal forms. In: "Problems of mechanics and mathematical physics". Nauka, 1976, 7-20 (Engl. transl.: Selecta Math. Sov. 1:1, 1981, 3-18).

67. On the theory of envelopes. Uspehy Math. Nauk, 1976, 31:3, 248-249.

68. Some unsolved problems of the singularity theory. In: Trans. of the seminar of S.L.Sobolev. Novosubirsk, 1976, 5-15.

69. Bifurcation of invariant manifolds of differential equations and structure of the neighborhood of an elliptic curve on a complex surface. FAA, 1975, 10:4, 1-12.

70. Wave front evolution and equivariant Morse Lemma. Comm Pure and Appl. Math. 1976, 29:6, 557-582.

71. Loss of stability of autooscillations near resonances and versal deformations of equivariant vector fields. FAA 1977, 11:2, 1-10.

72. Index of a singular point of a vector field, Petrovsky-Oleinik inequalities and mixed Hodge structures. FAA 1978, 12:1, 1-14.

73. Critical points of functions on manifolds with boundary, simple Lie groups B, C, F and singularities of evolutes. Rus. Math. Surv. 1978, 33:5, 91-105.

74. Additional chapters of the theory of ordinary differential equations. Moscow, Nauka, 1978, 304 p.

75. Some problems of theory of differential equations. In: Non-solved problems of mechanics and mathematics. Moscow State Univ. Press 1977, 3-9.

76. On the contemporary development of I.G.Petrovsky's works on topology of real algebraic manifolds. Uspehy Math. Nauk, 1977, 32:3, 215-216.

77. Indices of singular points of 1-forms on a manifold with a boundary, convolution of invariants of groups generated by reflections, and singular projections of smooth surfaces. Uspehy Math. Nauk, 1979, 34:2, 3-36.

78. On some problems in singularity theory. In: Geometry and Analysis, Papers dedicated to the Memory of V.K. Patodi. Proc. Indian Ac. Sci., 99:1, 1981, 1-9.

79. Real algebraic geometry (with O.A. Oleinik). Vestnik (Bulletin) Moscow State Univ., Ser. 1, 1979, No.6, 7-17.

80. Stable oscillations whose potential energy is harmonic on the space and periodic on the time. Appl. Math. Mach. 1979, 43:2, 360-363.

81. Catastrophe theory. Priroda 1979, No.10, 54-63.

82. Statistics of integer convex polyhedra. FAA 1980, 14:2, 1-3.

83. Lagrange and Legendre cobordisms. FAA 1980, 14:3, 1-13 and 14:4, 8-17.

84. Catastrophe theory. Moscow, Znanie Publishers, 1981, 64 p.

85. Large scale structure of the Universe I. General properties. One and two-dimensional models (with Ya.B. Zeldovich and S.F. Shandarin). Preprint Inst. Appl. Math. No.100, 1981, 32 p. (Engl. transl.: Geophys. Astrophys. Fluid Dynamics 1982, V. 20, 111-130).

86. Large scale structure of the Universe (with Ya.B. Zeldovich and S.F. Shandarin). Rus. Math. Surv. 1981, 36:3, 244-245.

87. Sweeping of the caustic by the cusps of moving front. Uspehy Math. Nauk, 1981, 36:4, 233.

88. Magnetic field in a moving conducting liquid (with Ya.B. Zeldovich, A.A. Rusmaikin and D.D. Sokolov). Uspehy Math. Nauk, 1981, 36:5, 220-221.

89. Magnetic field in a stationary flow with expansions in a Riemannian space (with Ya. B. Zeldovich, A.A. Rusmaikin and D.D. Sokolov). J. of Exp. and Theor. Phys. 1981, 81:6, 2052-2058.

90. Lagrange manifolds with singularities, asymptotical rays and unfurled swallowtail. FAA 1981, 15:4, 1-14.

91. Asymptotical rays in symplectic and contact geometry. Uspehy Math. Nauk, 1982, 37:2, 182-183.

92. Singularities of differentiable maps I. Classification of critical points, caustics and wave fronts (with A.N. Varchenko and S.M. Gusein-Zade). Moscow, Nauka, 1982, 304 p.

93. Stationary magnetic field in a periodic flow (with Ya.B. Zeldovich, A.A. Rusmaikin, D.D. Sokolov). Doklady AN USSR 1982, 266:6, 1357-1358.

94. Singularities of Legendre varieties, of evolvents and of fronts at an obstacle. Ergodic Theory and Dyn. Systems, 1982, v.2, 301-309.

95. On the Newtonian attraction of a multitude of dust-like particles. Uspehy Math. Nauk, 1982, 37:4, 125.

96. Surgeries of singularities of potential flows in a collisionless media and metamorphoses of caustics in a three dimensional space. Trans. of the I.G. Petrovsky seminar, 1982, vol.8, 21-57.

97. Some notes on the antidynamo theorem. Vestnik (Bulletin) Mosk. State Univ., Ser.1, 1982, No.6, 50-57.

98. On the Newtonian potential oh hyperbolic layers. Memoirs of Tbilisi Univ. 1982, vol. 232-233, 23-28 (Engl. transl.: Selecta Math. Sov., 1985, 4:2, 103-106).

99. Increase of a magnetic field in a three-dimensional flow of a noncondensable fluid (with E.I. Korkina) Vestnik (Bull.) Mosc. State Univ., Ser.1, 1983, No.3, 43-46.

100. Evolution of a magnetic field under the action of translaton and difusion. Uspehy Math. Nauk, 1983, 38:2, 226-227.

101. Singularities of systems of rays. Uspehy Math. Nauk, 1983, 38:2, 77-147.

102. Notes on the perturbation theory for the problems of Matieu type. Uspehy Math. Nauk, 38:4, 1983, 189-203.

103. Singularities in the variational calculus. Contemp. Probl. of Math. 1983, v.22, 3-55 (Engl. transl.: J. Soviet Math.)

104. Singularities, bifurcations and catastrophes. Uspehy Phys. Nauk (Soviet Phys. Uspehy), 1983, 141:4, 569-590.

105. Catastrophe theory, extended 2-d edition. Mosc. State Univ. Press, 1983, 80 p.

106. Geometrical methods in the theory of ordinary differential equations. Springer, New York a.o., 1983, 334 p.

107. Magnetic analogues of the Newton's and Ivory's theorems. Uspehy Math. Nauk, 1983, 38:5, 145-146.

108. Some algebro-geometrical aspects of the Newton attraction theory. Progress in Math., Vol.36, Birkhauser, Basel, 1983, 1-3.

109. Some open problems in the theory of singularities. Proc. of Symposia in Pure Math., Vol.40 Part 1, 1983, p.57-69.

110. Singularities of functions, wave fronts, caustics and multidimensional integrals (with A.N. Varchenko, A.B. Givental and A.G. Khovansky). Mathematical Physics Reviews, Vol.4, 1984, 1-92.

111. Catastrophe theory. Springer, Berlin, 1984, 79 p.

112. Singularities of differentiable maps II. Monodromy and asymptotics of integrals (with A.N. Varchenko and S.M. Gusein-Zade). Moscow, Nauka, 1984, 336 p.

113. Vanishing inflexions. FAA 1984, 18:2, 51-52.

114. Some remarks on the elliptic coordinates. Notes of LOMI Scientific Seminars. 1984, v.133, 38-50.

115. Singularities in the variational calculus. Uspehy Math. Nauk, 1984, 39:5,256.

116. On the evolution of magnetic field under the action of translation and diffusion. In.: Some problems of contemporary analysis, Mosc. State Univ. Press, 1984, 8-21.

117. Reversible systems. In: Nonlinear and Turbulent Processes. Gordon and Breach, New York 1984, 1161-1174.

118. Exponential dispersion of trajectories and its hydrodynamical applications. In: N.E. Kotchin and Development of Mechanics. Moscow, Nauka, 1984, 185-193.

119. English translation of No.92. Birkhauser, Boston a.o. 1985, 1-385.

120. Singularities of Ray Systems. Proc. Intern. Congr. Math., August 16-24, 1983, Warszawa, Vol. 1, 27-49.

121. Ordinary differential equations. 3-d edition, revized and expanded. Moscow, Nauka, 1984, 1-272.

122. Ordinary differential equations (with Yu.S. Il'ashenko). Current Probl. Math. VINITI, Fundamental Directions, Vol.1. Moscow, VINITI, 1985, 7-149 (Engl. transl. in: Springer, Encycl. of Math. Sciences, Vol.1)

123. Period maps and Poisson structures. Uspehy Math. Nauk, 1985, 40:5, 236.

124. Sturm theorems and symplectic geometry. FAA, 19:4, 1985, 1-10.

125. Superpositions. In: A.N.Kolmogorov, Selected Works, Mechanics and Mathematics. Moscow, Nauka, 1985, 444-451.

126. Classical mechanics. In: A.N.Kolmogorov, Selected Works, Moscow, Nauka, 1985, 433-444.

127. Implicit differential equations, contact structures and relaxation oscillations. Uspehy Math. Nauk, 40:5 (1985), 188.

128. Mathematical aspects of classical and Celestial Mechanics (with V.V. Kozlov and A.I. Neistadt). Moscow, VINITI, 1985. (Engl. transl.: Springer, Encycl. of Math. Sciences, Vol.3).

129. Symplectic geometry (with A.B. Givental). Current Probl. in Math. VINITI, Fundam. Directions, Vol.4. Moscow, VINITI, 1985, 7-139. (Engl. transl. in: Springer, Encycl. Math. Sciences, Vol.4)

130. On some nonlinear problems. In: Crafoord prize in mathematics, 1982. Crafoord lectures, The Royal Swedish Ac. of Sci., 1986, 1-7.

131. Catastrophe Theory. Second revized and expanded edition. Springer, Berlin, 1986, 108 pages.

132. Hyperbolic polynomials and Vandermond maps. FAA 20:2, 1986, 52-53.

133. Singularities of boundaries of spaces of differential equations. Uspehy Math. Nauk, 41:4, 1986, 152-154.

134. First steps of symplectic topology. Rus. Math. Surv., 41:6, 1986, 3-18.

135. Catastrophe theory and new possibilities of the application of mathematics. In: Mathematization of the contemporary science. Moscow, 1986, 81-87.

136. Bifurcation theory (with V.C. Afraimovitch, Yu.S. Il'ashenko and L.P. Shil'nikov). Current probl. of Math. VINITI, Fund. Directions, Vol.5. Moscow, VINITI, 1986, 5-218. (Engl. transl. in: Springer, Berlin a.o., Encycl. of Math. Sciences, Vol. 5)

137. Catastrophe theory. The same issue as for No.136, 219-277.

138. Oscillations and bifurcations in reversible systems (with M.B. Sevrjuk). In: Nonlinear Phenomena in plasma physics and Hydrodynamics, ed.: R.Z. Sagdeev. Mir Publishers, 1986, 31-64.

139. French translation of No. 92 and No. 112. Moscow, Mir Publishers, 1986.

140. 300-th Anniversary of the mathematical natural philosophy and celestial mechanics. Priroda, 1987, No.8(864), 5-15.

141. Hungarian edition of No.46. Budapest, 1987.

142. Quasicrystalls, Penrose partitions, Markov partitions, stochastic web and singularity theory. Uspehy Math. Nauk, 42:4, 1987, 139.

143. Second Kepler's law and topology of Abelian integrals (according to I.Newton). Kvant, 1987, No.12, 17-21.

144. Convex hulls and encreasing of productivity of systems in the pulsatory load. Siberian Math. J., 28:4, 1987, 29-31.

145. Contact structure, relaxational oscillations and singular points of implicit differential equations. In: Geometry and singularity theory in nonlinear equations. Voronezh, 1987, 3-8.

146. Portuguese transl. of No.59. Mir Publishers, 1987

147. Topological proof of the trnscendence of Abelian integrals in the Newtons "Principia". Histor.-Math. Investigations, XXXI, Moscow, Nauka, 1989, 7-17.

148. Ramified covering CP2 ? S4 , hyperbolicity and projective topology. Siberian Math. J., 29:5, 1988, 36-47.

149. Notes on the Poisson structures on the plane and other powers of the volume forms. Trans. of the I.G.Petrovsky seminar, No.12, 1987, 37-46.

150. German transl. of No. 72. Birkhauser, Basel, 1987, 320 p.

151. Mathematics with a human face. Priroda 1988, No.3, 117-119.

152. On surfaces, defined by hyperbolic equations. Matem Zametki (Math. Notices) 1988, 44:1, 3-17.

153. Remarks on quasicrystallic symmetries. Physica D, nonlinear phenomena. 1988, 33:(1-3), 21-25.

154. On some problems in symplectic topology. In: Topology and Geometry. Rohlin Seminar. O.Ya.Viro (Ed.) Springer Lecture Notes Math., 1346 (1988), 1-5.

155. On the interior scattering of waves, defined by hyperbolic variational principles. J. of Geometry and Physics 1988. Vol.V, No.3, 305-315.

156. Bifurcations and singularities in mathematics and mechanics. In: Theoretical and Applied Mechanics XVII IUTAM. Congress, Grenoble, 1988. Elsevier, 1989, 1-25.

157. German transl. of No.59. VEB Deutscher Verlag der Wissenschaften DDR, 520, 1988.

158. Singularities I. Local and Global Theory (with V.A. Vassil'ev, V.V. Gorjunov and O.V. Ljashko). Moscow, VINITI, 1988, 1-256 (to be transl. by Springer as Vol.6 of Encycl. Math. Sciences).

159. Singularities II. Classification and Applications (with V.A. Vassil'ev, V.V. Gorjunov and O.V. Ljashko). Moscow, VINITI, 1989, 1-256 (to be transl. by Springer as Vol. 39 of Encycl. Math. Sci.).

160. Dynamics of intersections. In: Analysis, et cetera. Research papers Published in Honor of Jurgen Moser's 60-th Birthday. Eds. P.Rabinovitz, E. Zehnder. Acad. Press. San Diego, 1990, 77-84.

161. A-graded algebras with 3 generators. Comm. Pure Appl. Math., 42, 1989, 993-1000.

162. Engl. transl. of No.112. Birkhauser, Boston a.o., 492 p.

163. Some words on Andrei Nikolaevich Kolmogorov. Uspehy Math. Nauk, 43:6, 1988, 34.

164. A.N.Kolmogorov in the reminiscences of pupils. Kvant, 1988, No.11-12, 34.

165. A.N.Kolmogorov, obituary. Physics today. 42:10, 1989, 148-150.

166. Contact structure, relaxation oscillations and singular points of implicit differential equations, in: Springer Lecture Notes Math., 1334, 173-179; 1988.

167. Spaces of functions with mild singularities. FAA, 23:3, 1988, 1-10.

168. Teoria delle catastrofi. Bollati Boringhieri, Torino, 1990, 146 p.

169. Newton's principia read 300 Years later (with V.A. Vassil'ev). Notices of the Amer. Math. Soc., 1989, 36:9, 1148-1154 and 37:2 (1990), 144.

170. Some unsolved problems of theory of differential equations and mathematical physics. Uspehy Math. Nauk, 1989, 44:4, 191-192.

171. Contact geometry: the geometrical method of Gibbs thermodynamics. AMS, 1990.

172. Singularity theory and its applications. Lezioni Fermiane. Academia Nazionale Dei Lincei. Scuola Normale Superiore. Pisa, 1990.

173. Catastrophe theory. Nauka i zhish'n (Science and life) 1989 No.10, 12-19.

174. Contact geometry and wave propagation. Monographie No.34 Enseign. Math., 1989, 56 pages.

175. Ten problems. In: Singularity theory and its applications. Adv. in Soviet math. Vol. 1. AMS, 1990, 1-8.

176. One hundred problems. MFTI, Moscow, 1989.

177. Bernoulli-Euler updown numbers associated with function singularities, their combinatorics and arithmetics. Duke Math. J., 63:2, 1991, 537-555.

178. Dynamics of complexity of intersections. Boletim da Sociedade Brasiliera de Mathematica 21:1, 1990, 1-10.

179. Huygens and Barrow, Newton and Hooke, Birkhauser, Basel a.o., 1990, 118 p.

180. Mathematical trivium. Uspehy Math Nauk. 46:1, 1991, 225-232.

181. Catastrophe theory. 3-d edition, extended. Moscow, Nauka, 1990, 128 p.

182. 3-d edition of No.59, extended. Moscow, Nauka, 1988. 472 p.

183. Singularities of Caustics and Wave Fronts. Kluwer, 1990.

184. Topological and ergodic properties of a closed differential 1-form, Func. Anal. Appl. 25:2 (1991),1-12.

185. Meanders. Kvant, 1991, no. 3, pp. 11-14.

186. Majoration of Milnor numbers of intersections in holomorphic dynamical systems, preprint 652 Utrecht Univ., April 1991, 1-9 (Topological Methods in Modern Mathematics, Publish or Perish 1992).

187. Springer numbers and morsification spaces, prprint 658 Utrecht Univ., April 1991, pp. 1-18. (J. Alg. Geom. 1:2, 1992)

188. Calculus of snakes, Uspehi Math. Nauk., 47, v. 2, 1992.

189. Problems on singularities and dynamical systems, Progress in Sov. Math., Chapman and Hall, 1992.

190. Topological methods in hydrodynamics (with B.A. Khesin). Annual Reviews in Fluid Dynamics, 24, 1992.

191. Mathematical trivium - II. Uspehy Math Nauk, 48:1, 1993, 211-222.

192. Bounds for Milnor numbers of intersections in holomorphic dynamical systems. In: Topological methods in Modern Mathematics (Stony Brook, NY, 1991). - Houston, TX: Publish or Perish, 1993, 379--390.

193. Sur les proprietes topologiques des projections lagrangiennes en geometrie symplectique des caustiques. -- CEREMADE, Universite Paris-Dauphine. Cahiers de Mathematiques de la Decision, 9320, 14.06.93, 9 p.

194. On the topological properties of Legendrian projections in the contact geometry of wave fronts. Algebra and Analysis, 1994, 6(3), 1--16 (in Russian).

195. Topological Invariants of Plane Curves and Caustics. Dean Jacqueline B. Lewis Memorial Lectures, Rutgers University. -- Providence, RI: Amer. Math. Soc., 1994, VIII+60 p. (University Lecture Series, 5).

196. Topological classification of real trigonometric polynomials and cyclic serpents polyhedron. In: Arnold--Gelfand Mathematical Seminars. Boston: Birkhauser, 1996.

197. Toronto lectures, June 1997
Lecture 1: From Hilbert's Superposition Problem to Dynamical Systems.
Lecture 2: Symplectization, Complexification and Mathematical Trinities.
Lecture 3: Topological Problems in Wave Propagation Theory and Topological Economy Principle in Algebraic Geometry.
(available at http://www.botik.ru/~duzhin/arnold/arn-papers.html).

198. On the problem of realization of a given Gaussian curvature function. Preprint, 12-Feb-98. (http://www.botik.ru/~duzhin/arnold/Gaussian.tex.gz).

Opening ceremony of the MCCME–IUM building: V. Arnold speaking,
A. Gonchar, Yu. Ilyashenko, A. Muzhykantski (seated left to right). (EMS
Newsletter, March 2010.)

In the Steklov Mathematical Institute of the RAS: a group of founders of the IUM

Academicians V.I.Arnold (chairman of the board of trustees, to the left) and S.P.Novikov.

SINGULARITIES OF FUNCTIONS, WAVE FRONTS, CAUSTICS AND MULTIDIMENSIONAL INTEGRALS

V. I. ARNOL'D, A. N. VARCHENKO, A. B. GIVENTAL' AND A. G. KHOVANSKII

Moscow State University, Moscow

Abstract

The theory of singularities is a grand-scale generalization of the study of maxima and minima of functions. In recent years a whole arsenal of methods has been developed in this mathematical theory, enabling one to give exhaustive answers to questions of the structure of singularities of wave fronts, caustics, and integrals of rapidly oscillating functions by the method stationary phase, saddle-point integrals with coalescing saddle points.

These methods also apply to the theory of Legendre transformations, to the bifurcation of equilibrium positions and oscillatory regimes of dynamical systems which depend on parameters, to optimization theory, calculus of variations, geometry of Riemannian manifolds, etc.

In this paper we discuss the basic ideas, concepts, and methods of the theory of singularities. The authors have taken measures to separate the conceptual aspects of the paper from the technical details; on the other hand, the text contains concrete results and tables needed for most applications.

From the point of view of mathematics, the success of the theory is based on the connection discovered in recent years between geometrical optics and the theory of asymptotic behavior of integrals and entities seemingly remote from them: simple Lie algebras and the theory of crystallographic groups generated by reflections.

The classification of the simplest singularities of caustics and wave fronts turns out to be related to the classification of compact Lie groups ($A_k = SU_{k+1}$, $D_k = SO_{2k}$, E_6, E_7, E_8), and also to the classification of regular polyhedra in three-dimensional space [i.e., the theory of the discrete subgroups of the group $SU(2)$].

In Section 1 we classify the singularities of systems of rays and fronts in geometrical optics: it is based on the theory of bifurcation of critical points in families of functions that depend on parameters. In Section 2 this information is used for studying short-wave asymptotics in the neighborhood of singularities of caustics. The calculation of the asymptotics of integrals reduces to the stereometry of certain convex polyhedra with integer vertices—the so-called Newton polyhedra of singularities (the vertices of the Newton polyhedra are the powers of monomials appearing in the Taylor series of the singularity). The Newton polyhedra also contain significant information about the topological invariants of manifolds given by algebraic or analytic equations (about the number of handles of Riemann surfaces, the number of solutions of polynomial systems of equations, etc.). Formulas for calculating these invariants are given in Section 3.

1

Section 1 was written mainly by A. B. Givental', Section 2 by A. N. Varchenko, and Section 3 by A. G. Khovanskii.
The authors are grateful to S. P. Novikov, who proposed that they write this survey.

Contents

§1   Singularities of Functions, Caustics, and Wave Fronts

The theory of singularities is concerned with the classification of degeneracies of objects of different types. In this section we shall study the singularities of functions, caustics, wave fronts, and their metamorphoses. A complete classification according to problems of interest would be a limitless task. The guiding thread we shall use through this labyrinth is the *principle of general position*: one should study only those singularities that cannot be eliminated by a slight deformation of the object. This principle is well illustrated by the singularities of level lines on a topographical map which shows peaks, dips and saddles. Upon closer inspection the ridges always turn out to be chains of mountain peaks and saddles.

For further information about the questions discussed in this section and related ones, we refer the reader to (Refs. 14, 12, 9, 57).

1.1 Examples

1.1.1   Wave Fronts and Caustics in the Plane   The propagation of disturbances in a medium can be described by means of a system of rays and by using wave fronts. Suppose, for example, that a disturbance propagates towards the interior of an ellipse in the plane with

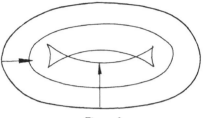

Figure 1

unit velocity. If we mark a segment of length t along the interior normals of the ellipse, we obtain a parallel to the ellipse, an equidistant wave front (Fig. 1). Although the initial front is smooth, after some time the parallel front develops singularities—cusps (the local equation of the front being $y^2 = x^3$ in some curvilinear coordinate system) and self-intersections. These singularities are *stable* (i.e., they cannot be eliminated by a slight change of the initial front). It turns out that all other singularities of the front in a plane (for example, the singularity at the center of a contracting circle) break up into these standard forms when there is a slight shift of the initial front.

A wave front moving in the plane traces out a certain surface in (three-dimensional) space–time. It appears that this surface can be regarded as a wave front (*grand front*) in space–time. The metamorphoses that the moving wave front experiences are identical to the metamorphoses of the sections of the grand front by isochrones (planes with $t = $ const) of space–time. The metamorphosis for the example of the ellipse is shown in Fig. 2. In this case the grand front singularity is called a "swallowtail." The "swallowtail" can be defined as the surface formed in x, y, z space by all the tangent lines to the curve $x = a^2$, $y = a^3$, $z = a^4$—its cuspidal edge. The metamorphosis of Fig. 2 exhausts all generic metamorphoses of wave fronts in the plane.

The envelope of rays normal to the front in the plane is called a *caustic* (see Fig. 3a). The light is concentrated at the points of the caustic. It is easy to show that the cusps on the moving front fill the caustic (Fig. 3b). Thus the caustic is obtained by projecting the cuspidal edge of the grand front from space–time onto ordinary space (Fig. 4). The singularities of the caustic in general position in the plane are exhausted by the cusp points and by the self-intersection points.

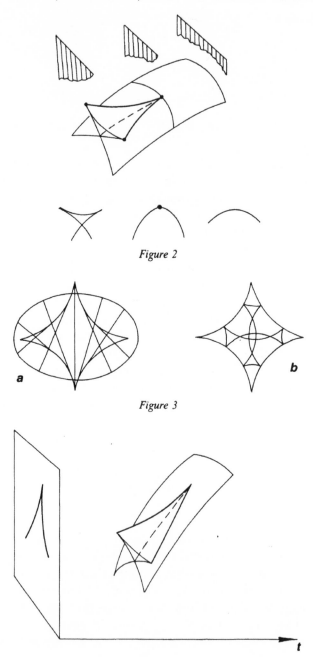

Figure 2

Figure 3

Figure 4

1.1.2 Singularities of a Function of a Single Variable Let $y = f(x)$ be a smooth (i.e., infinitely differentiable) function. A point in the domain of definition of the function is said to be *singular* if the derivative vanishes there. The value of the function at the singular point is called a *critical value*. If the second derivative of the function does not vanish at the singular point, the point is said to be *nondegenerate* or a type-A_1 point. In the neighborhood of a singular point of type A_1 the function can, by a smooth change of the variable, be brought to the form $y = \pm z^2 + \text{const}$. Similarly, if $f'(x_0) = f''(x_0) = 0$, while $f'''(x_0) \neq 0$, then the function can be brought to the form $y = z^3 + \text{const}$ in the neighborhood of the point x_0 (type-A_2 point); if $f'(x_0) = f''(x_0) = f'''(x_0) = 0$, while $f^{IV}(x_0) \neq 0$, the function in the neighborhood of the point x_0 can take the form $y = \pm z^4 + \text{const}$ (type-A_3 point), etc.

A function in general position has only nondegenerate singular points. Degeneracies are eliminated by arbitrarily small changes of the function (here and below a small change refers to the function and all its derivatives). Degenerate singularities appear when we consider families of functions. Thus, in the family $y = x^3 - tx$ of functions* of x with parameter t, for zero value of the parameter t there is a singular point of type A_2 at $x = 0$ (Fig. 5), and any neighboring family has the same singularity for some small value of

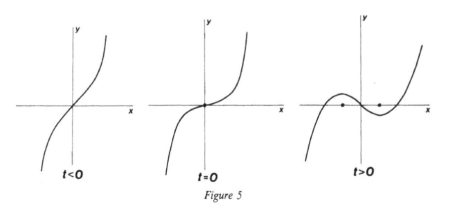

$t<0$ $t=0$ $t>0$

Figure 5

---

*By a family of functions we mean a function of two arguments, one of which is assumed to be a parameter. The arguments themselves may be points of multidimensional spaces. We shall sometimes call the parameter space the base of the family.

the parameter. In general one-parameter families of functions, A_2 singularities appear only for individual parameter values; in general two-parameter families of functions, A_2 singularities appear for parameter values that form a curve in the parameter plane, while one finds type-A_3 singular points for individual parameter values.

Functions in general position have no singular points with critical values that are *previously assigned* (e.g., zeros): such points are removed by adding a small constant to the function. A one-parameter family of functions in general position, in the neighborhood of a singular type-A_1 point with zero critical value can, by a small change of parameter λ, and a smooth change of variable that is a smooth function of the parameter, be brought to the form $y = \pm z^2 + \lambda$. The general two-parameter family of functions can be brought by these same operations to the form $y = z^3 + \lambda_1 z + \lambda_2$ in the neighborhood of a type A_2 point with critical value zero, and the general three-parameter family, in the neighborhood of a type-A_3 point with critical value zero, can be brought to the form $y = \pm z^4 + \lambda_1 z^2 + \lambda_2 z + \lambda_3$. The families of functions introduced here are said to be *versal deformations* of singular points of types A_1, A_2, A_3, respectively. The set of points in the parameter space of the versal deformation to which there correspond functions with zero critical values, is called the *bifurcation diagram of zeros*. For A_1 the bifurcation diagram of zeros consists of the single point $\lambda = 0$. For A_2, from the condition $y' = 0$ we find: $\lambda_1 = -3z^2$, while from the condition $y = 0$ we get $\lambda_2 = -z^3 - \lambda_1 z = 2z^3$; i.e., the bifurcation diagram of zeros is a semicubic parabola $4\lambda_1^3 + 27\lambda_2^2 = 0$ with cusp at $\lambda_1 = \lambda_2 = 0$ (Fig. 6). The bifurcation diagram of zeros for a singular type-A_3 point is a "swallowtail" (Fig. 7a) in the space of polynomials $z^4 + \lambda_1 z^2 + \lambda_2 z + \lambda_3$, consisting of polynomials with multiple roots. The projection of the singular points of the bifurcation diagram of zeros along the axis of the free term of the versal deformation onto the plane of the other parameters is called the *bifurcation diagram of functions*. For A_3 the bifurcation diagram of functions (Fig. 7b) consists of a semicubic parabola (functions $y = z^4 + \lambda_1 z^2 + \lambda_2 z$, with a singular type-$A_2$ point) and a ray (functions with two singular type-A_1 points with identical critical values). The shaded values of the parameters correspond to functions with three singular points, and the unshaded values correspond to those with one singular point. We see that the

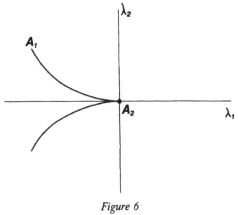

Figure 6

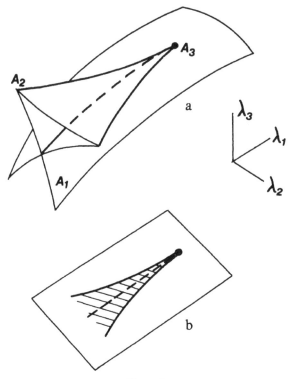

Figure 7

bifurcation diagrams carry considerable information about the breakup of singular points.

1.1.3   Modality, Transversality, and Codimension   For greater clarification of the concepts used by us—stability, families, versal deformations, and bifurcation diagrams—we use the following geometric interpretation. We shall picture the objects to be classified (functions, for example) as points of a space (which is actually infinite-dimensional). The classification is a decomposition of the space of all objects (Fig. 8). Objects in general position occupy almost all the space. Degenerate objects occupy the surfaces of separation (the coordinate crossing in Fig. 8) between the regions of objects in general position (the four quadrants). A *stable* object is one for which all the nearby objects are in the same class as it is in (points in quadrants II, III and IV). In Fig. 8 the classes in the first quadrant depend on a continuous parameter (the *modulus*)—the slope angle. The *modality* of an object is the largest number of parameters on

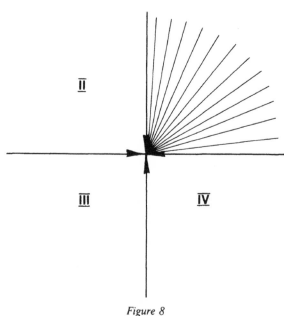

Figure 8

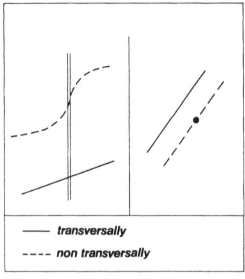

Figure 9

which the classes depend, which intersect a small neighborhood of this object in the space of objects. In Fig. 8 the points of the first quadrant, the positive coordinate semiaxes, and the coordinate origin are unimodal, while the other points are 0-modal. A 0-modal object, i.e., one whose neighborhood includes a finite number of classes, is said to be *simple*. The example of a discrete classification, one in which all objects are simple, gives the classification of the singular points of analytic functions of a single variable: a class of the singular point is determined by the order of the first nonzero term in the Taylor series of the function at that point. A *family* of objects is represented by a curve, surface, etc. in the space of all objects. Families in general position intersect the simple classes *transversally* (i.e., at a nonzero angle, Fig. 9). In particular, in one-parameter families one finds that only those classes cannot be eliminated which are represented as surfaces of *codimension* one (i.e., selected by one equation in the space of objects); in two-parameter families the classes that cannot be eliminated are simple classes of codimension two (two equations, see Fig. 10), etc. In studying finite-parameter families we can ignore sets of infinite codimension in the object space

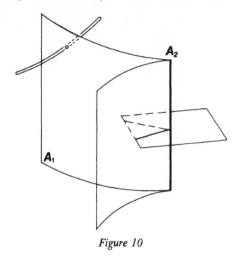

Figure 10

(for example, the set of functions with null Taylor series in the space of smooth functions).

1.2   Critical Points of Functions

1.2.1   Basic Concepts   A *singular* or *critical* point of a smooth function is a point in the domain of definition of the function at which its differential vanishes. In considering the behavior of a function in the neighborhood of a critical point, we shall say that two functions are equivalent (more precisely, *R*-equivalent) if they go over into each other under a suitable, smooth change of variables in the neighborhoods of the critical points.* A critical point of a function is said to be nondegenerate if the quadratic form of the second differential of the function is nondegenerate at that point. A function in the neighborhood of a nondegenerate critical point is equivalent to the quadratic part of its Taylor series (M. Morse), i.e., in a certain coordinate system it has the form

$$\pm x_1^2 \pm \cdots \pm x_n^2 + \text{const}$$

---

*The term "*R*-equivalent" comes from the word "right," since only the changes in the variables on the right side of the function symbol in the expression $y = f(x)$ are permitted.

A smooth function in general position has only nondegenerate singular points. Degenerate singular points are eliminated by an arbitrarily small distortion of the function. (Here and below a small change also applies to higher derivatives).

After a small shift (deformation) of the function, a degenerate singular point splits into several nondegenerate points (see Fig. 5). If their number is bounded for any deformation, such a degenerate singular point is said to have *finite multiplicity*. In the neighborhood of a critical point with finite multiplicity a smooth function is equivalent to its Taylor polynomial of sufficiently high order (J. C. Tougeron). If we consider not only real critical points but also complex ones, we find that under deformations of this polynomial the number of singular points into which a degenerate singular point splits is independent of the manner of deformation—so long as all are nondegenerate (just as an n-fold root of a polynomial in one variable splits into n simple roots, which are generally complex). This number is called the *multiplicity* or *Milnor number* of the singularity.

If at a critical point of a function of n variables the quadratic form of the second differential has rank r (i.e., reduces to a sum of r squares), then the function in the neighborhood of this point has, in a certain coordinate system, the form

$$\text{const} \pm x_1^2 \pm \cdots \pm x_r^2 + \varphi(x_{r+1}, \ldots, x_n),$$

where the Taylor series of the function φ at the origin begins with terms of degree three. The number $n - r$ is called the *corank* of the singular point. Functions that may depend on different numbers of variables are said to be *stably equivalent* in the neighborhood of 0 if they become equivalent after adding to them nondegenerate quadratic forms in a suitable number of new variables. Stably equivalent critical points of functions of the same number of variables, with zero second differential, are equivalent.

1.2.2 Simple Singularities and Their Miniversal Deformations The classes of stably equivalent simple critical points of smooth functions with zero critical values form the infinite series $A_\mu^\pm$ ($\mu \geqslant 1$), $D_\mu^\pm$ ($\mu \geqslant 4$) and the classes $E_6^\pm$, E_7, and E_8 that abut each other, as

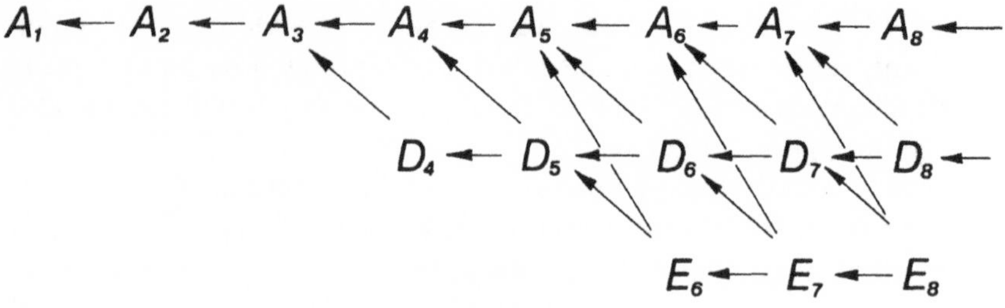

Figure 11

shown in Fig. 11. Below we enumerate examples of the classes of functions with critical point 0:

$$A_\mu^\pm : \pm x^{\mu+1}, \qquad D_\mu^\pm : x^2y \pm y^{\mu-1},$$

$$E_6^\pm : x^3 \pm y^4, \qquad E_7 : x^3 + xy^3, \qquad E_8 : x^3 + y^5.$$

The class A_{2k}^+ coincides with the class A_{2k}^-, and the singular points of the classes A_1^+ and A_1^- are stably equivalent. With these exceptions, the classes enumerated are all different. The subscript gives the multiplicity of the singular point. In families of functions in general position with a number of parameters $l \leqslant 6$, among the singular points with zero critical value one finds only simple singular points of multiplicity $\mu \leqslant l$.

Deformations of a critical point in which it remains unremovable (i.e., every neighboring deformation contains an equivalent critical point) are called *versal* (more precisely, *R*-versal). Among versal deformations, the *miniversal* deformations have the smallest number of parameters. Infinite-multiplicity critical points which form a set of infinite codimension can be eliminated from the finite-parameter families of functions. The codimension of an *R*-equivalence class of a finite-multiplicity singular point in the space of all singular points of functions is equal to its multiplicity. A miniversal deformation is represented as a transversal addition to the equivalence class (see Fig. 10), so that the dimension of the base of a miniversal deformation of a singular point is equal to its multiplicity.

For simple singular points of the functions described above one can choose as miniversal the following deformations:

$$A_\mu^\pm : x^{\mu+1} + \lambda_1 x^{\mu-1} + \lambda_2 x^{\mu-2} + \cdots + \lambda_{\mu-1} x + \lambda_\mu,$$

$$D_\mu^\pm : x^2 y \pm y^{\mu-1} + \lambda_1 y^{\mu-2} + \cdots + \lambda_{\mu-1} + \lambda_\mu x,$$

$$E_6^\pm : x^3 \pm y^4 + \lambda_1 y^2 x + \lambda_2 xy + \lambda_3 x + \lambda_4 y^2 + \lambda_5 y + \lambda_6,$$

$$E_7 : x^3 + xy^3 + \lambda_1 y^4 + \lambda_2 y^3 + \lambda_3 y^2 + \lambda_4 y + \lambda_5 + \lambda_6 xy + \lambda_7 x,$$

$$E_8 : x^3 + y^5 + \lambda_1 y^3 x + \lambda_2 y^2 x + \lambda_3 yx + \lambda_4 x + \lambda_5 y^3$$
$$+ \lambda_6 y^2 + \lambda_7 y + \lambda_8.$$

For an arbitrary function f with μ-fold singularity at the point 0, a miniversal deformation is constructed as follows. Just like the remainders from the division of integers by some fixed integer, we consider the "remainders from division" of functions by all the partial derivatives of the function f. The remainders form a μ-dimensional space. This means that there exist μ functions $\varphi_1, \ldots, \varphi_\mu$, such that any function defined in the neighborhood of the point 0_x is uniquely representable in the form $\sum c_j \varphi_j + \psi$, where c_j are constants, and ψ is "divisible" by $\partial f / \partial x$: $\psi = \sum h_i \partial f / \partial x_i$ for certain functions $h_1, \ldots, h_n$. For example, for $f = x^3/3$ each function of x has a remainder after division by $\partial f / \partial x = x^2$ of the form $c_1 \cdot 1 + c_2 \cdot x$. A deformation $f + \lambda_1 \varphi_1 + \cdots + \lambda_\mu \varphi_\mu$ is miniversal for the singular point considered. In Fig. 12 we show how one calculates the "remainders" for the function $f = x^3 + xy^3$ with singularity E_7. The monomial $x^m y^l$ corresponds to the point (m, l). The shaded monomials have zero "remainders," a nonzero linear combination of monomials joined by lines is divisible by $\partial f / \partial x$, and the circles mark the basis of "remainders" $\varphi_1, \ldots, \varphi_7$. We are convinced that the Milnor number of singularity E_7 is equal to 7.

An arbitrary deformation of the critical point is obtained from a miniversal deformation by the following construction.

By means of a smooth mapping $t \mapsto \lambda(t)$ of the space T into the base of the family of functions $\mathcal{F}(x, \lambda)$, we can *induce* the new family:

$$G(x, \lambda) = \mathcal{F}\big[x, \lambda(t)\big].$$

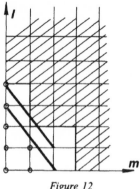

Figure 12

Two families of functions $G(x,\lambda)$ and $\mathscr{F}(x,\lambda)$ are said to be *R-equivalent* if one goes into the other after a change of parameter $\Lambda(\lambda)$ and a change of variables $X_\lambda(x)$, which smoothly depend on the parameter λ: $G(x,\lambda) = \mathscr{F}(X_\lambda(x), \Lambda(\lambda))$.

Versality theorem (J. Mather). Any deformation of a finite-multiplicity critical point is equivalent to a deformation induced from a versal deformation. Versal deformations with the same number of parameters are equivalent to each other. Deformations obtained from a versal deformation by adding new parameters on which it is not explicitly dependent are also versal.

1.2.3 Bifurcations and Dynkin Diagrams For the majority of points λ of the base of miniversal deformation of a critical point, the corresponding function f_λ has no singular points with zero critical value. The set of points of the base of miniversal deformation for which the function f_λ has such singular points, is called the *bifurcation diagram of zeros*.

Example: Miniversal deformation of A_μ singularity is the family of polynomials with highest coefficient unity and zero sum of roots; its bifurcation diagram of zeros, which consists of polynomials having multiple roots, is called a generalized swallowtail.

In par. 1 we saw the adjacencies of strata of the bifurcation diagram of zeros carry significant information about the breakups of the initial singular point (see Fig. 7). To enumerate all the strata of bifurcation diagrams of simple singularities, it is convenient to use Dynkin diagrams. The Dynkin diagram of a singularity of multiplicity

μ is a collection of μ points joined by edges, having the following form:

$$A_\mu \bullet—\bullet—\bullet \cdots \bullet—\bullet—\bullet \qquad D_\mu \bullet—\bullet \cdots \bullet—\bullet\!\!<^{\textstyle\bullet}_{\textstyle\bullet}$$

$$E_6 \bullet—\bullet—\!\!\!|\!\!\!—\bullet—\bullet \qquad E_7 \bullet—\bullet—\!\!\!|\!\!\!—\bullet—\bullet—\bullet \qquad E_8 \bullet—\bullet—\!\!\!|\!\!\!—\bullet—\bullet—\bullet—\bullet$$

The Dynkin diagrams of classes of simple singularities, whose designations differ only in a $\pm$ sign, are identical. Identification of these classes is actually done if we consider, instead of the classification of singular points of smooth functions, the classification of singular points of complex analytic functions. For example, the functions $x^3 + y^4$ and $x^3 - y^4$, with singular points of types E_6^+ and E_6^-, go over into each other under the substitution $y \to \sqrt{-1}\, y$. All the preceding concepts and results carry over word for word to this case. The classes of stable equivalence of simple singular points of complex-analytic functions with zero critical values form two infinite series A_μ and D_μ and three classes E_μ. The miniversal deformations of complex singularities are the same as above; we need only consider the variables x, y and the parameters λ to be complex. The strata of the complex bifurcation diagram of zeros and the decay of a simple singular point in the complex domain are calculated as follows. The decay due to deformation of a simple critical point into several simple critical points, lying on the *zero* level of the deformed function will be associated with a set of Dynkin diagrams corresponding to the equivalence classes of these critical points. One can show (E. V. Brieskorn) that such a correspondence is a one-to-one correspondence between the possible decays of the initial simple singularity on the zero level surface of the function and the *subgraphs* of its Dynkin diagram, obtained by removing a certain number of vertices and the edges that emerge from them. The complex dimension of a stratum of the bifurcation diagram of zeros, whose points correspond to a given decay, is equal to the number of vertices removed. Inclusion of one subdiagram in another corresponds to the adjacency of the corresponding strata. The decays of a D_4 singularity, the complex dimensions of the strata and their adjacencies are enumerated in Fig. 13.

The bifurcation diagrams of zeros of versal deformations are obtained from the bifurcation diagrams of miniversal deformations by a

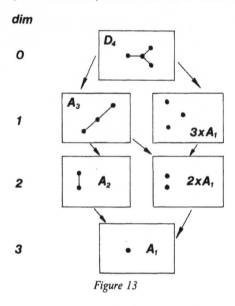

Figure 13

Cartesian product on the space of auxiliary parameters. The decomposition into versal deformations is the same as into miniversal, but the dimensions of the corresponding strata differ in the number of superfluous parameters, so that the codimension of a stratum corresponding to a given decay is independent of the choice of versal deformation and is equal to its codimension in the space of all functions.

1.2.4 Unimodal Critical Points   One has now obtained the classification of singularities of functions with a number of moduli ≤ 2 (V. I. Arnol'd). At the same time, one obtains the classification of singular points of multiplicity ≤ 16 and codimension ≤ 10 (for the definition see par. 1.2.5). Below we give the normal forms of unimodal singular points with zero critical value. The parameter a is the modulus.

(a)   Parabolic singularities (simple singularities are also called elliptic):

$$P_8 : x^3 + ax^2z \pm xz^2 + y^2z, \qquad a^2 \neq 4, \quad \text{if} \quad +$$

$$X_9 : \pm x^4 + ax^2y^2 \pm y^4, \qquad a^2 \neq 4, \quad \text{if} \quad + + \quad \text{or} \quad - -$$

$$J_{10} : x^3 + ax^2y^2 \pm xy^4, \qquad a^2 \neq 4, \quad \text{if} \quad +$$

Table I

Designation	Normal Form	Restrictions
J_{10+k} $(T_{2,3,6+k})$	$x^3 \pm x^2 y^2 + a y^{6+k}$	$a \neq 0, k > 0$
X_{g+k} $(T_{2,4,4+k})$	$\pm x^4 \pm x^2 y^2 + a y^{4+k}$	$a \neq 0, k > 0$
$Y_{r,s}$ $(T_{2,r,s})$	$\pm x^2 y^2 \pm x^r + a y^8$	$a \neq 0, r, s > 4$
$\tilde{Y}_r$ $(\tilde{T}_{2,r,r})$	$\pm (x^2 + y^2)^2 + a x^r$	$a \neq 0, r > 4$
P_{8+k} $(T_{3,3,3+k})$	$x^3 \pm x^2 z + y^2 z + a z^{3+k}$	$a \neq 0, k > 0$
$R_{l,m}$ $(T_{3,l,m})$	$x(x^2 + yz) \pm y^l \pm a z^m$	$a \neq 0, m \geqslant l > 4$
$\tilde{R}_m$ $(\tilde{T}_{3,m,m})$	$x(x^2 \pm y^2 \pm z^2) + a y^m$	$a \neq 0, m > 4$
$T_{p,q,r}$	$a xyz \pm x^p \pm y^9 \pm z^r$	$a \neq 0, 1/p + 1/q + 1/r < 1$
$\tilde{T}_{p,m}$ $(\tilde{T}_{p,m,m})$	$x(y^2 + z^2) \pm x^p + a y^m$	$a \neq 0, 1/p + 2/m < 1$

(b) Hyperbolic singularities, given in Table I.
(c) Exceptional singularities:

$$K_{12} : x^3 + y^7 + axy^5 \quad K_{13} : x^3 + xy^5 + ay^8 \quad K_{14} : x^8 \pm y^3 + axy^6$$

$$Z_{11} : x^3 y + y^5 + axy^4 \quad Z_{12} : x^3 y + xy^4 + ay^6 \quad Z_{13} : x^3 y \pm y^6 + axy^5$$

$$Q_{10} : x^2 z + y^3 \pm z^4 + ayz^3 \qquad Q_{11} : \pm x^2 z + y^3 + yz^3 + az^5$$

$$Q_{12} : x^2 z + y^3 \pm z^5 + ayz^4$$

$$W_{12} : \pm x^4 + y^5 + ax^2 y^3 \qquad W_{13} : \pm x^4 + xy^4 + ay^6$$

$$U_{12} : x(x^2 \pm y^2) \pm z^4 + axyz^2$$

$$S_{11} : x^2 z + yz^2 \pm y^4 + ay^3 z \qquad S_{12} : x^2 z + yz^2 + xy^3 + ay^5$$

In the complex domain the functions in this list, which differ in the $\pm$ sign, are R-equivalent. Representatives of the complex hyperbolic classes can be chosen in the single form:

$$T_{p,q,r} : x^p + y^q + z^r + axyz, \qquad a \neq 0, \qquad \frac{1}{p} + \frac{1}{q} + \frac{1}{r} < 1.$$

Thus the unimodal classes of stable equivalence of singular points of complex-analytic functions form three parabolic one-parameter families P_8, X_9, J_{10}, a three-index series of hyperbolic one-parameter families $T_{p,q,r}$ and 14 exceptional one-parameter families. Some of the adjacencies between unimodal classes are shown in Fig. 14.

1.2.5 R^+-equivalence and Bifurcation Diagrams of Functions So far we have dealt with critical points with fixed (zero) critical value. In

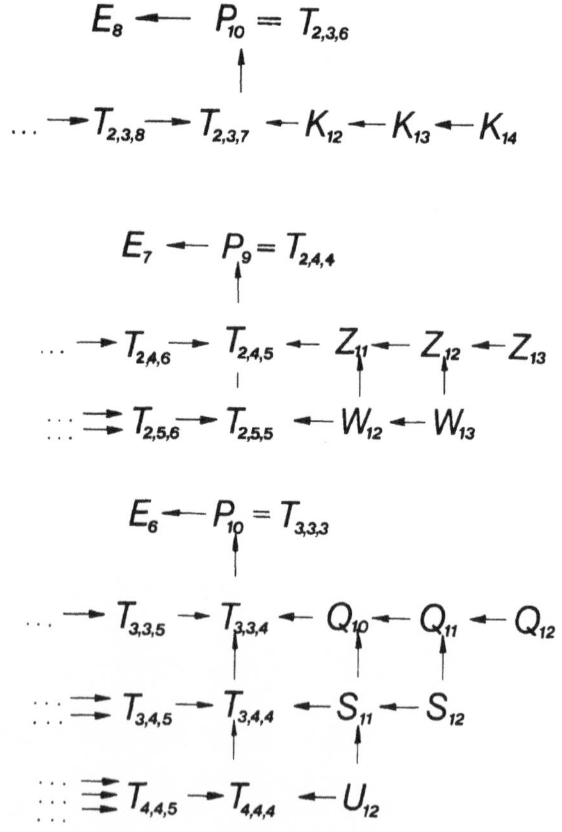

Figure 14

order to remove this restriction we introduce a definition: two functions in the neighborhood of a singular point are said to be R^+-
equivalent if they become R-equivalent after adding a suitable constant to one of them. Of course, the R^+-classification of singularities
coincides with the R-classification introduced in pars. 1.2.2 and 1.2.4.
The whole theory is constructed by analogy with the preceding
discussion.

The distinction between the R^+ and R theories begins with the
introduction of the concept of versal deformation. An R^+-miniversal
deformation is constructed from a set of remainders *that vanish at the
singular point*, or, in other words, are obtained from an R-miniversal
by dropping the free term. In the basis of R^+-miniversal deformations we distinguish the *bifurcation diagram of functions*, consisting of
two strata, called *caustics* and *stratum of intersections*. To the first
stratum belong all points λ of the base of R^+-versal deformations for
which the function $\mathscr{F}(\cdot,\lambda)$ of the variable x (obtained for a fixed
parameter value λ) has degenerate critical points; to the second
stratum belong all those points of the basis of R^+-versal deformations
for which the function $\mathscr{F}(\cdot,\lambda)$ has multiple critical values. The
bifurcation diagram of functions is obtained by projection of the set
of singular points of the bifurcation diagram of zeros along the axis of
the free term of the R-miniversal deformation (see Fig. 7). The inverse
image of a point not lying on the bifurcation diagram of the function
is a line parallel to this axis, and intersects the bifurcation diagram of
zeros in a finite number of points, equal to the number of (nondegenerate) critical points of the corresponding function. For the A_μ
singularity the caustic is the same as the bifurcation diagram of zeros
of the $A_{\mu-1}$ singularity, i.e., a generalized swallowtail. This corresponds to a semicubical parabola in Fig. 7. Caustics of classes $D_4^\pm$
lying in three-dimensional space, have special names. They are shown
in Fig. 15. Characteristic sections of the bifurcation diagrams of
functions for A_4 and $D_4^\pm$ are shown in Fig. 16.

Two families of functions are said to be R^+-*equivalent* if they
become R-equivalent after adding to one of them a constant that
depends smoothly on a parameter. As in R-theory, the versality
theorem holds: any deformation of a singular point of a function is
R^+-equivalent to one induced from an R^+-versal deformation.

Families of functions in general position, depending on $l \leqslant 5$ parameters, have only simple singularities of multiplicity no higher than

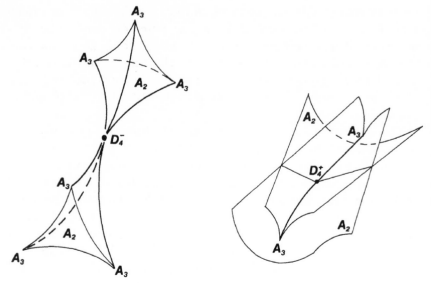

Figure 15

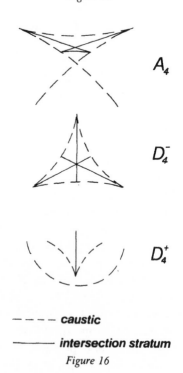

Figure 16

$l + 1$. Such families are stable—all neighboring families of functions are R^+-equivalent. For families with a larger number of parameters this is no longer the case. Let us consider as an example the R^+-miniversal deformation of a P_8 singularity (see par. 1.2.4):

$$x^3 + y^3 + z^3 + \lambda_1 xyz + \lambda_2 xy + \lambda_3 yz + \lambda_4 zx + \lambda_5 x + \lambda_6 y + \lambda_7 z.$$

In the family obtained from the miniversal for $\lambda \equiv 0$, the singularity of the function $x^3 + y^3 + z^3$ is removable, but in any neighboring family, e.g., for $\lambda \equiv \text{const} \neq 0$, one finds, instead of it, a singularity of $x^3 + y^3 + z^3 + \lambda_1 xyz$ having the same multiplicity, and differing only in the value of the family modulus λ_1.

We define the *codimension* of a singular point as the codimension in the space of functions with critical point 0 of the set of functions with singular point lying either in the same R^+-equivalence class as the given singularity or in a class differing from the first only in the values of the moduli. One can show (A. M. Gabrielov) that the codimension, multiplicity μ, and modality m of a singular point of a function are connected by the relation $c = \mu - m - 1$. Thus, for singularities of type P_8 the codimension is 6, and these singularities are nonremovable from 6-parameter families of functions.

1.2.6 Singularities of a Hypersurface and V-equivalence of Functions
A *hypersurface* in a space is a set with codimension one; it is given by a single equation $f(x) = 0$. If a point of the hypersurface is not a singular point of the function defining the surface, then by the implicit function theorem, the hypersurface is smooth in the neighborhood of the point. The singular points of the hypersurface with $f = 0$ are those singularities of the function f with zero critical value. But the hypersurface can also be given by a different equation $g(x) \cdot f(x) = 0$, if the function g does not vanish anywhere. We say that the critical points of functions with zero critical values are *V-equivalent* if after a suitable change of variables these functions will differ only by a nonvanishing factor.* The classification of the singular points of hypersurfaces reduces to the *V*-classification of the critical points of functions.

---

* *V*-equivalence is an equivalence of the zero level varieties. *V* is the first letter of the word "variety."

In considering families of hypersurfaces in general position, we again go to the concept of versal deformation. Thus, the general 3-parameter family of plane curves in the neighborhood of a singular point of the curve $y^2 = x^4$ can be brought to the form $y^2 = x^4 + \lambda_1 x^2 + \lambda_2 x + \lambda_3$. The V-miniversal deformation of a finite-multiplicity singular point of the function f has the form $f + \lambda_1 \varphi_1 + \cdots + \lambda_k \varphi_k$, where $\varphi_1, \ldots, \varphi_k$ are "remainders from division" of functions defined in the neighborhood of the critical point by the function f and its partial derivatives, i.e., they are obtained by dropping those terms of the R-miniversal deformation which have the same remainders after division by the partial derivatives as functions that are multiples of f.

A function is said to be *quasihomogeneous* if its arguments $x_1, \ldots, x_n$ can be assigned positive (quasi) degree or weights such that all monomials appearing in the Taylor series of this function assume the same quasidegree. The normal forms of simple, parabolic, and exceptional (for $a = 0$) singularities introduced in pars. 1.2.2 and 1.2.4 have this property. For example, the function $f(x, y) = x^2 y + y^{\mu - 1}$ becomes quasihomogeneous if we set $\deg y = 1/(\mu - 1)$, $\deg x = (\mu - 2)/(2\mu - 2)$. K. Saito has shown that a function in the neighborhood of a singular point is divisible by its partial derivatives if and only if it is R-equivalent to a quasihomogeneous function [for example: $f(x, y) = (\mu - 2)/(2\mu - 2) \cdot x \partial f/\partial x + 1/(\mu - 1) \cdot y \partial f/\partial y$]. Thus, V-versal deformations of quasihomogeneous singularities coincide with R-versal deformations. One can show that in the nonquasihomogeneous case the bifurcation diagram of zeros of an R-versal deformation is obtained from the bifurcation diagram of zeros of a V-versal deformation by taking a Cartesian product with the space of missing parameters.

Two families of functions are said to be V-*equivalent* if they become R-equivalent after multiplication of one of them by a nonvanishing function that depends smoothly on the parameters.

The versality theorem is valid also in V-theory: any deformation of a singular point is V-equivalent to one induced from a V-versal deformation.

In families of hypersurfaces with $l \leqslant 6$ parameters, only simple singular points with multiplicity $\mu \leqslant l$ are unremovable. General families of hypersurfaces with no more than 6 parameters are stable.

The V-classification of singular points of functions differs from the R-classification. For example, in the classes of simple functions there is an identification $A_\mu^+ \sim A_\mu^-$, $D_\mu^+ \sim D_\mu^-$, $E_6^+ \sim E_6^-$. Thus the simple classes of stable V-equivalence of singular points of smooth functions form two infinite series A_μ and D_μ and three classes E_μ. Richer examples are given by the exceptional singularities. For example, in the family K_{12}, for nonzero values of the modulus a the functions $x^3 + y^7 + axy^5$ are V-equivalent to each other (prove!) and are V-nonequivalent to the quasihomogeneous function $x^3 + y^7$.

1.2.7 Boundary Singularities By a submanifold *with a boundary* we mean a half-space selected from the space R^{n+1} with coordinates $x, y_1, \ldots, y_n$ by the inequality $x \geqslant 0$. A point $x = 0$ of the boundary is said to be a *boundary singular* point for a smooth function of the variables x, y, if the level surface of the function that passes through it is not transverse to the boundary (in other words, this point is critical, in the usual sense, for the restriction of the function to the boundary). Two boundary singular points of a function are said to be *equivalent* if the functions in the neighborhood of these points go over into each other under a smooth change of variables that takes the boundary and the half-space $x \geqslant 0$ into themselves. A *simple* boundary singular point with zero critical value is stably equivalent to the singular point 0 of one of the following functions (V. I. Arnol'd):

$$A_\mu', D_\mu', E_\mu' : f(x, y_1, y_2) = \pm x + g(y_1, y_2);$$

$$B_\mu : f(x) = \pm x^\mu, \quad \mu \geqslant 2; \quad C_\mu : f(x, y) = xy \pm y^\mu, \quad \mu \geqslant 3;$$

$$F_4 : f(x, y) = \pm x^2 + y^3,$$

where the function g has a simple singular point (of type A_μ, D_μ, or E_μ) at the origin. Miniversal deformations of the simple singular points have the form

$$B_\mu : \pm x^\mu + \lambda_1 x^{\mu-1} + \cdots + \lambda_\mu, \quad C_\mu : xy \pm y^\mu + \lambda_1 y^{\mu-1} + \cdots + \lambda_\mu,$$

$$F_4 : \pm x^2 + y^3 + \lambda_1 xy + \lambda_2 x + \lambda_3 y + \lambda_4, \quad \tilde{A}_\mu, \tilde{D}_\mu, \tilde{E}_\mu : x + \mathcal{F}(y, \lambda),$$

B

where $\mathscr{F}(y,\lambda)$ is a miniversal deformation of the corresponding singularity of A_μ, D_μ, E_μ.

Functions in general position on a manifold with boundary have only *nondegenerate* singular points (of type A_1 outside the boundary or type A_1' on the boundary). The corresponding *bifurcation diagram of zeros* for the boundary singularity consists of two hypersurfaces: on one of them lie those points λ of the base of miniversal deformation for which the manifold of the zero level of the function $\mathscr{F}(\cdot,\lambda)$ of the miniversal family is singular, on the other lie those points λ for which this manifold is nontransversal to the boundary. Thus, for the miniversal deformation $\mathscr{F}(x,\lambda_1,\lambda_2) = \pm x^2 + \lambda_1 x + \lambda_2$ of a B_2 singularity, the first hypersurface is given by the conditions $x \geqslant 0$, $\partial\mathscr{F}/\partial x = \pm 2x + \lambda_1 = 0$, $\mathscr{F} = \pm x^2 + \lambda_1 x + \lambda_2 = 0$, from which we find $\lambda_2 = \pm\lambda_1^2/4$, $\pm\lambda_1 \leqslant 0$ while the second hypersurface is given by the conditions $x = \mathscr{F} = 0$, from which $\lambda_2 = 0$, i.e., the bifurcation diagram of zeros of B_2 is a semiparabola, tangent to the line. For the B_3 and C_3 singularities, the bifurcation diagrams of zeros are shown in Fig. 17. One can show that the complex bifurcation diagram of zeros looks the same for C_μ as for B_μ, but the first and second hypersurfaces interchange roles. The adjacencies of the simple classes of singular points of functions on a manifold with boundary are shown in Fig. 18.

In the families of functions on manifolds with boundary with $l \leqslant 3$ parameters, the only unremovable singularities are those simple ones with multiplicity $\mu \leqslant l + 1$. Nonsimple singular points are unremovable, starting with 4-parameter families of functions. As in par. 1.2.5, we can define R^+-versal deformations of boundary singularities. Pictures of caustics of certain boundary singularities are shown in par. 2.1.5.

The theory of singularities of functions on a manifold with boundary includes the theory of functions on a manifold without a boundary, since the function can also have singularities outside the boundary. On the other hand, the substitution $x = z^2$ converts a function on a manifold with boundary $x = 0$ into a function on a manifold without a boundary, even in the variable z; a change of variables (x, y), which takes the boundary and the half-space $x \geqslant 0$ into themselves, goes over into the change of variables (z, y) taking symmetric points into symmetric points. Therefore, the theory of

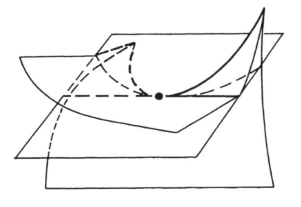

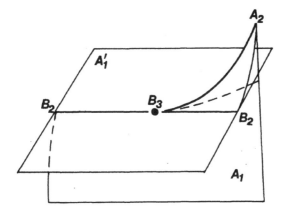

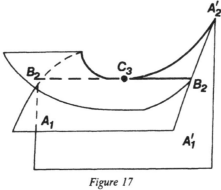

Figure 17

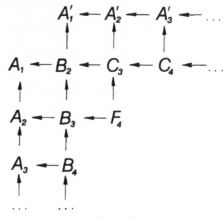

Figure 18

boundary singularities of functions is equivalent to the theory of singularities of even functions (with the corresponding understanding of equivalence).

1.3   Singularities of Wave Fronts

1.3.1   Legendre Mappings   A wave front in the plane is a curve with singular points. It turns out that it can be realized as the projection of a *smooth* curve in 3-space onto the plane. Singularities of the front arise where this curve is poorly placed relative to the projection direction. For example, consider the cusp of the wave front $y = \pm x^{3/2}$ (Fig. 19). In the space with coordinates x, y, p, consider the curve $y = \pm x^{3/2}$, $p = dy/dx = \pm \frac{3}{2} x^{1/2}$. It is smooth(!), since it is

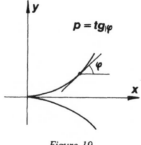

Figure 19

given by the equations $x = 4/9p^2$, $y = 8/27p^3$. The point (x, y, p) of three-dimensional space determines a contact element of the plane—the tangent line with equation $dy = p\,dx$, attached at the point (x, y). Our smooth curve is called a *lift* of the semicubical parabola into the space of contact elements.

We note that the projection of a smooth curve in general position from space onto the plane has as singularities only self-intersections and no cusps. The singularities of wave fronts are singularities of the special class of smooth mappings of n-dimensional manifolds into $(n + 1)$-dimensional ones.

A *contact element* of the space $\mathbb{R}^{n+1}$ is a tangent hyperplane attached at any point. It is given by the equation $dy = p_1 dq_1 + \cdots + p_n dq_n$, where $y, q_1, \ldots, q_n$ is a coordinate system in $\mathbb{R}^{n+1}$, while $p_1, \ldots, p_n$ are the angular coefficients of the hyperplane. The manifold E of all contact elements has dimension $2n + 1$. In the space tangent to E at the point (q, y, p) the equation $dy = p\,dq$ gives a $2n$-dimensional *contact hyperplane*. The field of contact hyperplanes on E actually is independent of the choice of coordinates (q, y) in $\mathbb{R}^{n+1}$.*

The *lift* of a smooth hypersurface in $\mathbb{R}^{n+1}$ consists of all the contact elements tangent to it and has dimension n. Such a hypersurface is locally the graph of a function $y = f(q)$. Its lift is given by the auxiliary equation $p = \partial f/\partial q$, and consequently at each of its points is tangent to the contact hyperplane $dy = p\,dq$. Smooth n-dimensional submanifolds of the space of contact elements, tangent to the field of contact hyperplanes, are called *Legendre manifolds*. It is easy to verify that the fibres of the projection of the space of contact elements in $\mathbb{R}^{n+1}$, consisting of contact elements attached at a given point of the space $\mathbb{R}^{n+1}$, are Legendre fibres. The projection of the space of contact elements into the $(n + 1)$-dimensional space (basis), whose fibres are Legendre fibres, is said to be a *Legendre fibration*.

Suppose that we are given a Legendre submanifold in the space E of a Legendre fibration. Its projection on the base of the fibration is called a *Legendre mapping*.

---

*This field is defined intrinsically as follows: The velocity vector of a moving contact element lies in a hyperplane of the field if the velocity vector of the point of application of the contact element belongs to the contact element.

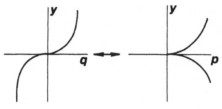

Figure 20

Examples of Legendre mappings: (1) In addition to the space $\mathbb{R}^{n+1}$, we consider the $n + 1$-dimensional space of all hyperplanes in $\mathbb{R}^{n+1}$ (not necessarily passing through the origin). A mapping that associates to a point of a smooth hypersurface in $\mathbb{R}^{n+1}$ the tangent hyperplane at that point is a Legendre mapping. Its image—a hypersurface in the space of all hyperplanes, is called the *Legendre transform* of the original hypersurface and consists of all hyperplanes tangent to it (see Fig. 20).

(2) The mapping that associates to a point of a hypersurface in Euclidean space the end of the exterior normal vector of length t, is a Legendre mapping. This mapping is called a *normal mapping* and its image is called the *equidistant* of the original hypersurface.

The singularities of a Legendre transformation and of the normal mapping of a generic hypersurface are the same as those of a Legendre mapping in general position.

The motion of the wave front in space corresponds to the motion of its lift—a Legendre manifold in the space of contact elements. If the initial front is smooth, then its Legendre manifold remains smooth, even when the front acquires singularities. Thus, *the singularities of wave fronts are the singularities of projections of Legendre submanifolds of the space of contact elements under a Legendre mapping.*

Under a diffeomorphism (smooth mapping) of the space $\mathbb{R}^{n+1}$, the tangent vectors are also transformed, and consequently also the contact elements, so that we may speak of the *lift of a diffeomorphism* of the space $\mathbb{R}^{n+1}$ to a diffeomorphism of the space of contact elements. We shall say that two Legendre mappings are *equivalent* if their projected Legendre manifolds go over into each other under a lift diffeomorphism of the space $\mathbb{R}^{n+1}$. It is obvious that when there is equivalence of Legendre mappings, their fronts also go over into each other.

1.3.2 Generating Families of Hypersurfaces A Legendre manifold, which projects nonsingularly on the base of the Legendre fibration, is given by its front $y = f(q)$ as follows:

$$y = f(q), \qquad p = \partial f / \partial q.$$

It turns out that an arbitrary Legendre mapping can be given analogously by means of a *generating family of hypersurfaces*. It is constructed as follows. Let I be a subset of the set of indices $\{1, \ldots, n\}$, such that the projection $(q, y, p) \mapsto (q_J, y, p_I)$ (where J is the set of residual indices) is nonsingular for the given Legendre submanifold in the neighborhood of the given point (it always exists). The equality

$$d(y - p_I q_I) = - q_I dp_I + p_J dq_J$$

shows that the projection $(q, y, p) \mapsto (- q_J, p_I, y - p_I q_I)$ of the space of contact elements is a Legendre fibration in which our Legendre submanifold has a nonsingular front. Suppose that $y - p_I q_I = f(p_I, q_J)$ is the equation of this front. Then the Legendre manifold is given by the conditions

$$y - p_I q_I = f(p_I, q_J), \qquad q_I = - \partial f / \partial p_I, \qquad p_J = \partial f / \partial q_J.$$

The family $y = \mathscr{F}(p_I, q) = f(p_I, q_J) + p_I q_I$ of hypersurfaces in p_I-space with parameters (q, y) is called a *generating family* of the Legendre mapping whose Legendre manifold is defined by the equations

$$y = \mathscr{F}(p_I, q), \qquad \partial \mathscr{F} / \partial p_I = 0, \qquad p_J = \partial \mathscr{F} / \partial q_J,$$

and whose front consists of those points of the parameter space for which the corresponding hypersurface of the family is singular, i.e., the front is the bifurcation diagram of zeros of the family $y = \mathscr{F}$! In particular, the wave front in general position is the bifurcation diagram of zeros of the family in general position.

Examples and remarks: (1) The family $p^3 + qp - y = 0$ determines the Legendre manifold $q = - 3p^2$, $y = - 2p^3$ and the front $27y^2 + 4q^3 = 0$.

(2)   The front of the family $p_1^4 + q_2 p_1^2 + q_1 p_1 - y = 0$ is the surface of a "swallowtail."

(3)   The generating family has a simple physical meaning. Suppose that λ is the point of observation, x the source point, and $\mathcal{F}(x,\lambda)$ the optical path length. Then the wave front at the time t is the set of points λ, for which there exists a point x of the source at which $\mathcal{F}_\lambda = t$ and $\partial \mathcal{F}_\lambda / \partial x = 0$.

THEOREM 1 (L. Hörmander, A. Weinstein)   Two Legendre mappings are equivalent if and only if their generating families are V-equivalent.

COROLLARIES   (1)   The stable singularities of wave fronts are exhausted by the singularities of bifurcation diagrams of zeros of V-versal deformations of functions and their transversal intersections.

(2)   Singularities in general position of wave fronts of dimension $n \leqslant 5$ in $n + 1$-dimensional space are stable, and are exhausted by the bifurcation diagrams of zeros of V-versal deformations of singularities of functions of types A_μ, D_μ, and E_μ with $\mu \leqslant n + 1$.*

(3)   The singularities of wave fronts in general position in three-dimensional space are exhausted by cuspidal edges (A_2) and by singularities of the swallowtail type (A_3).*

(4)   Wave fronts in general position of dimension $n \geqslant 6$ can be unstable, at least at individual points.

1.3.3   Metamorphoses of Wave Fronts   Let us consider a time-dependent wave front. The Cartesian product of spaces which contain the front along the time axis is called *space–time*, whose projection on the time axis is the time function, and the union of all instantaneous fronts is called a *grand front*. The grand front is the front of a certain Legendre mapping in space–time. Suppose that the instantaneous front at the time t is given by the generating family of functions $\mathcal{F}_t(x,\lambda)$ (where λ is a space point). Regarding t as a new parameter, we again determine the formula for the Legendre manifold in the space of contact elements of space–time. Its projection into space–time is the grand front.

---

*It is of course still possible to have transversal intersections of different branches of wave fronts.

We already know that the grand front of a family of fronts in general position, for low dimension, is the bifurcation diagram of zeros of a simple singularity of functions. Thus we must study the generic decompositions of bifurcation diagrams into the level sets of the time function.

Two functions (time functions) in the base of R-versal deformations of the critical point are said to be *equivalent* if they go into each other under a suitable diffeomorphism of the base, preserving the bifurcation diagram of zeros, and under addition to one of them of a suitable constant.

We choose as the miniversal deformations of simple quasihomogeneous singularities the deformations given in par. 1.2.2, so that λ_1 is the coefficient of the "remainder" of largest quasihomogeneous degree. A *special metamorphosis* of the front is one in which the space–time has coordinates $\lambda_1, \ldots, \lambda_\mu, \tau_1, \ldots, \tau_m$, the grand front is

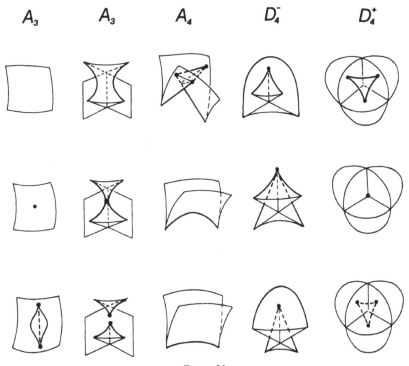

Figure 21

the direct product of the bifurcation diagram of zeros in the space of parameters λ on the parameter space τ, while the time function has the form

$$t = \tau_1 \quad \text{or} \quad t = \pm \lambda_1 \pm \tau_1^2 \pm \cdots \pm \tau_m^2 .$$

THEOREM 2 (V. I. Arnol'd) A generic metamorphosis of wave fronts in spaces of dimension $l \leqslant 5$ is locally equivalent to the special metamorphosis in which $\mu + m = l + 1$.

Under the special metamorphosis with time function $t = \tau_1$, the front is actually not changed. Metamorphosis of the second type for a front in three-dimensional space is shown in Fig. 21.

1.4   Caustics and their Metamorphosis

1.4.1   Geometry of Phase Space   A first-order partial differential equation, i.e., a relation $G(x, y, \partial y/\partial x) = 0$ between independent variables x, the function $y(x)$ and its partial derivatives, can be interpreted as a hypersurface $G(q, y, p) = 0$ in the space of contact elements equipped with the field of contact hyperplanes $dy = p\, dq$. The lifts of graphs of solutions of such an equation are the Legendre submanifolds of this hypersurface. An equation in which the required function y does not appear explicitly is called a Hamilton–Jacobi equation. The function G of the coordinates q of configuration space and the momenta p is the Hamiltonian of the mechanical system corresponding to this equation. Just as the mathematical apparatus for our study of wave fronts was the geometry of the space of contact elements, so in the theory of the singularities of caustics we shall deal with the geometry of phase space of Hamiltonian mechanics.

All the concepts of contact geometry have analogs in Hamiltonian mechanics. The role of the field of hyperplanes $dy = p/dq$ is taken by the *action form* $p\, dq$. The Legendre manifolds for projection along the y axis (Fig. 22) go over into *Lagrangian manifolds*—smooth submanifolds with dimension half that of the phase space, on which the action form is locally the differential of a function (in other words, the action integral $\oint p\, dq$ around any small contour is zero). For example,

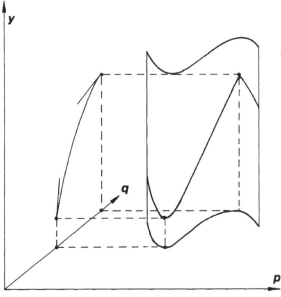

Figure 22

any curve in the phase plane is a Lagrangian curve. The fibration of phase space into Lagrangian fibres is called a *Lagrangian fibration*. An example of a Lagrangian fibration is the projection $(q, p) \mapsto q$ of phase space onto configuration space with fibres $q = $ const.

Suppose that there is a wave front moving in the configuration space. In the extended phase space (q, p, t) this motion corresponds to a "grand" Legendre manifold (t plays the role of y), and in the phase space (its projection) a Lagrangian manifold (Fig. 22). A subset in q-space marked by singularities of the moving front (the two points in Fig. 22) coincides with the critical values of the projection of the Lagrangian submanifold onto the basis of the Lagrangian fibration. We call such a projection a *Lagrangian mapping*, and the set of its critical points, a *caustic*.

1.4.2 Examples of Lagrangian Mappings (1) The mapping of a hypersurface of Euclidean space onto the unit sphere of that space, which associates to a point of the hypersurface the unit normal vector at that point, shifted to the coordinate origin, is called a *Gaussian*

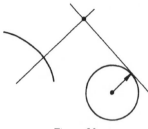

Figure 23

mapping. We represent the Gaussian mapping as the composition of two mappings: the first takes a point of the hypersurface into the point of intersection of the normal with the tangent space to the sphere that is orthogonal to it (Fig. 23), while the second consists of projecting the tangent space of the sphere at its point of application and is a Lagrangian fibration of the phase space of all tangent vectors to the sphere onto the configuration space. One can show that the image of the first mapping is a smooth Lagrangian submanifold of the phase space.

(2) The mapping that associates its terminus to a vector normal to a submanifold of Euclidean configuration space (the vector p, applied at the point q, corresponds to the point $p + q$) is called a *normal mapping.* The set of all normal vectors to the submanifold is a Lagrangian submanifold of the phase space of all vectors, tangent to the configuration space. The critical values of the normal mapping, which are the centers of curvature of the original submanifold, form the caustic of the family of its equidistants (see Fig. 3, in which the submanifold is an ellipse).

(3) Suppose that we are given a function f in Euclidean configuration space. The mapping that associates to a point of the configuration space the value of the gradient of the function at the point is called the *gradient mapping.* The submanifold of phase space consisting of all vectors tangent to the configuration space consisting of the points $(q, p = \partial f / \partial q)$ is a Lagrangian, since the action is the differential of the function f on it: $p\,dq = \partial f / \partial q\,dq = df$. The projection $(q, p) \mapsto p$ of the phase space onto p-space is a Lagrangian fibration.

The Gaussian, normal, and gradient mappings are special classes of mappings of manifolds of the same dimension. One can show that the typical singularities of these three classes coincide with the typical singularities of Lagrangian mappings and, generally speaking, differ from the typical singularities of general mappings of manifolds of the same dimension.

1.4.3 Generating Families of Functions Suppose that we are given a Lagrangian submanifold in the space of a Lagrangian fibration. The function y on this manifold, whose differential is the action $p\,dq$ (in the neighborhood of the particular point) determines the Legendre submanifold of the extended phase space with coordinates (q, y, p) and field of contact hyperplanes $dy = p\,dq$. The generating family of hypersurfaces $y = \mathscr{F}(p_I, q)$ of this Legendre manifold can be regarded as a family of graphs of the family of functions $\mathscr{F}(p_I, q)$. The family $\mathscr{F}$ is called the *generating family of functions* of the given Lagrangian mapping. In terms of its generating family, the Lagrangian manifold is defined by the equations:

$$\partial \mathscr{F}(p_I, q)/\partial p_I = 0, \qquad p_J = \partial \mathscr{F}/\partial q_J$$

(Here I and J are a decomposition of the indices $\{1, \ldots, n\}$, such that the functions (p_I, q_J) form a coordinate system in our Lagrangian manifold at the point considered), while the caustic consists of those points of the configuration space of parameters for which the functions of the family have degenerate singular points, i.e., it is exactly the caustic of the bifurcation diagram of the functions of the family $\mathscr{F}$.

Just as in the case of wave fronts, the physical meaning of the generating family $\mathscr{F}(x, q)$ is that it is the optical path length from the source point x to the observation point q.

We say that two Lagrangian mappings are *Lagrange equivalent* if there exists a diffeomorphism of the phase space that takes the fibres of the Lagrangian fibration into fibres that preserve the action integral over any closed contour and which maps the first Lagrangian submanifold onto the second. Since Lagrange equivalence takes the fibres of a Lagrangian fibration into fibres, it induces a diffeomorphism of the base of the fibration, taking one caustic into another. The converse is not true: from diffeomorphism of caustics it does not follow that the Lagrange mappings are Lagrange equivalent.

THEOREM 3 (L. Hörmander, A. Weinstein) Lagrange mappings are Lagrange equivalent if and only if their generating families of functions are R^+-equivalent.

COROLLARIES (1) The generating families of functions of stable local Lagrangian mappings are exhausted by the R^+-versal deformations of singularities of the functions.

(2) Singularities of caustics in general position in a space of dimension $n \leqslant 5$ are stable and exhausted by the caustics of R^+-versal deformations of singularities of functions such as A_μ, D_μ, and E_μ, with $\mu \leqslant n + 1$.*

(3) Singularities of caustics in general position in three-dimensional space are exhausted by the cuspidal edges (A_3) and by the singular points such as the "swallowtail," "pyramid," and "purse" $(A_4, D_4^-, D_4^+$, see Fig. 15).*

(4) Caustics in general position in a space of dimension $n \geqslant 6$ may be unstable, at least at individual points.

When the wave front moves, its cuspidal edges slip along the caustic. We obtain a decomposition of the caustic by isochrones. In Fig. 24 we show such a decomposition of the swallowtail (see Fig. 21,

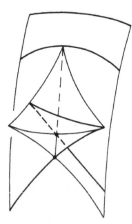

Figure 24

---

*Transverse crossings of different branches of caustics are also possible.

A_4). This decomposition of the swallowtail into curves differs from the decomposition of the "grand front" in Fig. 2 into instantaneous plane fronts: the cuspidal edges of the moving front do not have self-intersections, and pass twice through each point of self-intersection of the swallowtail, with a time interval of the order of $t^{5/2}$, where t is the distance to the vertex of the swallowtail.

1.4.4 Metamorphoses of Caustics Let us consider a medium of noninteracting and noncolliding particles, moving in configuration space by inertia with a velocity field $v(q)$. After a time t, the particle from point q goes to the point $q + t \cdot v(q)$. We obtain a family of mappings g_t of the configuration space into itself, which is a gradient mapping if the velocity field is potential:

$$v = \partial S/\partial q, \qquad g_t : q \mapsto \partial\big(q^2/2 + t \cdot S(q)\big)/\partial q.$$

Let us assume that the particles of the medium, as before, do not collide, but are moving in a potential force field, which possibly depends on the time. Since during the motion of a Hamiltonian system the action integral over a closed contour is conserved, the mapping through the time t, which is determined by the motion of our system, will be a Lagrangian mapping, provided that the initial velocity field is potential.

If the particles at the initial moment were distributed in space with a positive density, then the density will change as a result of the motion. The faster particles will begin to overtake the slower ones (Fig. 25) and, after the time t, the density ρ will become infinite at the points of the caustic of the mapping for time t (i.e., at the points of

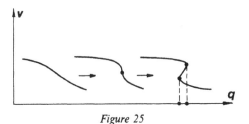

Figure 25

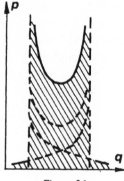

Figure 26

collision of infinitely close particles, Fig. 26). We thus arrive at the problem of describing the metamorphosis of the caustic in time.*

Considering the time as a new parameter of the generating family of functions $\mathscr{F}_t$ of the Lagrangian mapping, we get the generating family of a "grand" Lagrangian mapping of space-time into itself. The problem of describing the general metamorphoses of instantaneous caustics then reduces to the problem of describing the metamorphoses of sections of the "grand caustic" in general position in space–time by the level hypersurfaces of the time function. Two time functions are said to be *equivalent* if they go over into each other under some diffeomorphism of space–time that preserves the "grand caustic" and under a diffeomorphism of the time axis. A list of normal forms of time functions for cases when the grand caustic has singularities A_μ has been calculated by V. I. Arnol'd, and for D_μ by V. M. Zakaljukin. The grand caustic can be described as the set of critical values of the mapping $(x, \lambda, \tau) \mapsto (y = \partial \mathscr{F}/\partial x, \lambda, \tau)$, where in case A_μ

$$\mathscr{F} = \pm x^{\mu+1} + \lambda_1 x^{\mu-1} + \cdots + \lambda_{\mu-2} x^2, \qquad \lambda \in \mathbb{R}^{\mu-2}, \quad \tau \in \mathbb{R}^m,$$

---

*According to Ya. B. Zeldovich, at an early stage in the existence of the universe, the velocities of the particles formed a potential field, and the appearance of caustics during further motion of the particles is responsible for the present large-scale clumped structure of the universe (extreme nonuniformity in the distribution of the accretions of galaxies).

and in case D_μ

$$\mathcal{F} = x_1^2 x_2 \pm x_2^{\mu-1} + \lambda_1 x_2^{\mu-2} + \cdots + \lambda_{\mu-3} x_2^2, \qquad \lambda \in \mathbb{R}^{\mu-3}, \quad \tau \in \mathbb{R}^m$$

(where the dimension of the space–time is $\mu - 1 + m$).

THEOREM 4   (1) The time function t in general position can be reduced to the form

for $A_\mu : t = \tau_1$ or $t = \pm\lambda_1 \pm \tau_1^2 \pm \cdots \pm \tau_m^2$,

for $D_\mu : t = \tau_1$ or $t = \pm\lambda_1 + y_1 + a\lambda_2 \pm \tau_1^2 \pm \cdots \pm \tau_m^2$

by a caustics metamorphoses equivalence (for $\mu = 4$, $a\lambda_2$ must be replaced by ay_2, $y_k = \partial\mathcal{F}/\partial x_k$).
(2) In general one-parameter families of caustics in spaces of dimension $n \leqslant 4$, one meets only metamorphoses of types A_μ and D_μ with $\mu - 2 + m = n$.
In Figs. 27, 28 and 29 we show metamorphoses of caustics in general position in the plane and in space.
Returning to the consideration of a collisionless medium, we shall characterize the singularity of the caustic of the transformation after time t by the average density in an ϵ-neighborhood of the point we

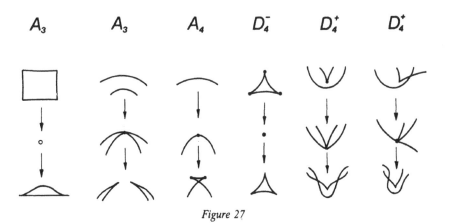

Figure 27

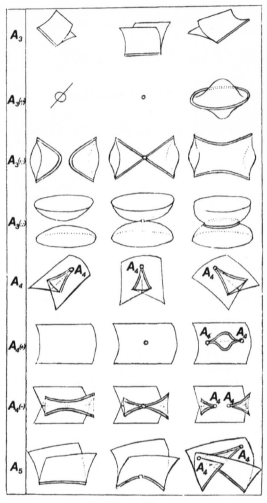

Figure 28

consider (the ratio of the mass in this neighborhood to its volume). At points of the caustic the average density tends to infinity, when the radius of the neighborhood ϵ goes to zero. In Table II we give the order of magnitude of the average density at different points of the caustic. Singularities A_5 and D_5 appear only at individual times.

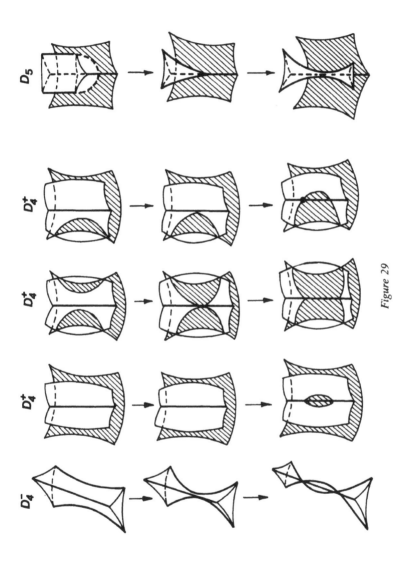

Figure 29

Table II

Caustic	Cusp	Swallowtail	Pyramid Purse	A_5	D_5
$\epsilon^{-1/2}$	$\epsilon^{-2/3}$	$\epsilon^{-3/4}$	ϵ^{-1}	$\epsilon^{-4/5}$	$\epsilon^{-1}\ln\epsilon^{-1}$

§2 Integrals of the Stationary Phase Method

This section is concerned with the study of the asymptotics of *oscillatory integrals*, i.e., integrals of the form

$$I(\tau) = \int_{\mathbb{R}^n} e^{i\tau f(x)}\varphi(x)\,dx_1 \ldots dx_n, \tag{1}$$

for large values of the real parameter τ. Here f and φ are smooth functions which are called, respectively, the *phase* and the *amplitude*. According to the principle of stationary phase, the main contribution to the asymptotic comes from the neighborhood of the critical points of the phase. In this section we discuss the application of the theory of singularities to the study of asymptotics, and present results of calculations of asymptotics in the simplest generic situations, and indicate formulas for calculating asymptotics.

2.1 Examples

2.1.1 Oscillation Integrals and Short-Wave Oscillations Problems in optics, acoustics, quantum mechanics, various branches of mathematics (theory of partial differential equations, probability theory, number theory) require analyzing oscillatory integrals for large values of the parameter.

Example: Consider a surface in space. Assume that each point of the surface radiates a spherical wave of fixed frequency and fixed wave length. Assume that the wavelength is small compared to the size of the surface and to the rate of change of the wave amplitude as we vary the point on the surface. We shall be interested in the overall oscillatory regime at space points outside the surface.

We select a point on the surface and describe the sum of the waves coming from a small neighborhood of this point to a fixed point of the space. This sum is determined by the manner in which the distance from the surface point to the space point varies for small changes of the surface point. If the straight line along which the wave propagates from the surface point to the space point is not perpendicular to the surface, then the arriving waves cancel out each other. If the line along which the wave propagates is perpendicular to the surface, then many waves will arrive at the space point from the selected point of the surface with approximately equal phases. The area of the points on the surface from which waves arrive with approximately equal phases is proportional to the wavelength (in fact, this area is proportional to the area of the set of points at which the modulus of a quadratic expression in the local coordinates is less than the wavelength). Consequently, the contribution of this point of the surface to the total oscillatory regime is proportional to the wavelength. It may turn out that the change in distance to the space point for small variations of the surface point has a higher than second order of smallness (this occurs when the quadratic form for the change in distance is degenerate). In this case the area of the set of points of the neighborhood, from which waves with approximately equal phase arrive, is even greater, and is proportional to the wavelength to a lower power than 1.

The total oscillation at the point y is given by the function

$$e^{2\pi i \omega t} \int_S \frac{e^{2\pi i \|x-y\|/\lambda} \varphi(x)}{\|x - y\|} \, dx$$

where t is the time, ω is the frequency, λ is the wavelength, φ is the amplitude, dx is the element of area of the surface, and S is the surface that radiates the wave. Thus the total oscillation is given by an oscillatory integral, in which the role of the large real parameter is played by a quantity inverse to the wavelength, while the phase is the distance function from the surface point to the fixed point in space. The preceding considerations show that the main contribution to the total oscillation (the oscillatory integral) comes from the neighborhood of critical points of the phase. If all the critical points of the phase are nondegenerate, the contribution to the total vibration from each of them is proportional to the wavelength. If the phase has degenerate critical points, then the contribution from their small neighborhoods is greater, the order of their contribution being proportional to some power of the wavelength smaller than 1. As a rule, the function on the surface, which is equal to the distance to the fixed space point, has only nondegenerate critical points. A space point is said to be *caustic* or *focal* if the function on the surface, equal to the distance to the space point, has a degenerate critical point. The caustic points form a new surface in space, called the *caustic*. At points of the caustic, the total oscillation has a nonstandard large value. If the surface radiates light waves, the caustic is a surface of points with nonstandard brightness. It can be seen on a wall illuminated by rays that are reflected from a concave surface, e.g., the surface of a cup.

Many examples of problems in which the need arises to study asymptotics of oscillating integrals are given in the papers of M. Berry and J. Nye, cited in the bibliography. See, for example, Ref. 46, where the photographs of caustics formed by rays of light refracted in a drop of water are shown.

2.1.2   The Principle of the Stationary Phase States   The main contribution to an oscillatory integral comes from the neighborhoods of critical points of the phase; if the phase of the oscillating integral has no critical points on a support of the amplitude, then as the parameter in the integral tends to infinity the integral will tend to zero faster than any power of the parameter.

If we assume that the integral is one-dimensional, we can integrate it by parts:

$$\int e^{i\tau f(x)}\varphi(x)\,dx = -1/(i\tau)\int e^{i\tau f(x)}(\varphi(x)/f'(x))'\,dx.$$

Repeating the integration a sufficient number of times, we obtain our assertion. The multidimensional case is easily reduced to the one-dimensional case.

2.1.3   Fresnel Integral   An oscillatory integral whose phase has only nondegenerate critical points is called a *Fresnel integral*.

Example: Consider an oscillatory integral with phase x^2. Figure 30 shows the graph $y = \cos(\tau x^2)\varphi(x)$ of the real part of the integrand.

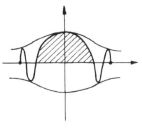

Figure 30

It is clear that for large values of the parameter τ the integral is proportional to the area under the first peak of the graph, i.e., it is proportional to $\varphi(0)\tau^{-1/2}$.

Consider the multiple Fresnel integral

$$\int_{\mathbb{R}^n} e^{i\tau f(x)}\varphi(x)\,dx_1\,\ldots\,dx_n.$$

Assume that the phase of the integral has a nondegenerate critical point at the origin and amplitude zero in the neighborhood of other critical points of the phase.

THEOREM 1 (see Ref. 10) As the parameter τ of the integral tends to $+\infty$, the integral is representable in the form

$$\varphi(0)(2\pi/\tau)^{n/2}\exp(i\tau f(0)$$

$$+ i\pi/4\,\text{Sign}\,f''_{xx}(0))|\det f''_{xx}(0)|^{-1/2} + O(\tau^{-n/2-1}).$$

where f''_{xx} is the matrix of second partial derivatives of the phase, Sign f''_{xx} is the signature* of this matrix.

Example: If $f = x_1^2 + \cdots + x_k^2 - x_{k+1}^2 - \cdots - x_n^2$, then the leading term in the asymptotic expression is

$$\varphi(0)(\pi/\tau)^{n/2}\exp(i\tau f(0) + i\pi(2k - n)/4).$$

---

*The *signature* of a nondegenerate symmetric matrix is the difference in the numbers of its positive and negative eigenvalues.

2.1.4   Caustics   In applications the phases and amplitudes of the oscillatory integrals, as a rule, depend on auxiliary parameters. Let us consider such an integral.

Assume that the phase, considered as a family of functions depending on parameters, is a family of functions in general position (see Section 1). In this case the integral is of Fresnel type for almost all values of the parameters, and at these values is of order $\tau^{-n/2}$.

Parameters for which the phase has a degenerate critical point form a hypersurface in the parameter space. This hypersurface is called a *caustic* (see par. 1.4). For caustic values of the parameters the order of magnitude of the integral is determined by the degenerate critical points of the phase.

2.1.5   Asymptotics of Oscillatory Integrals near a Caustic   Assume that for some selected value of the auxiliary parameters the phase of the oscillatory integral has a single critical point, and that the phase, considered as a family of functions depending on the parameters, is a family of functions in general position. In this case the caustic in the neighborhood of the selected parameter is said to be *elementary*.

Examples of elementary caustics, occurring for numbers of parameters equal to 2 and 3, are shown in Figs. 31 and 32. In these figures, we indicate near each part of the caustic the type of degenerate critical points of the phase occurring for these caustic values of the parameters. For example, $A_2 + A_2$ means that the phase has two critical points of type A_2, while the remaining critical points of the phase are nondegenerate. Each degenerate critical point of the phase contributes of order $\tau^{\beta-n/2}$ to the integral. The number β for critical points of types A_μ and D_μ is equal, respectively, to $(\mu - 1)/(2\mu - 2)$ and $(\mu - 2)/(2\mu - 2)$ (see, below, Theorem 7).

THEOREM 2 (see Refs. 12 and 13)   For phases depending in a general way on two or three parameters, each elementary caustic is locally diffeomorphic to one of the caustics shown in Figs. 31 and 32.

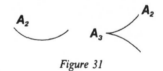

Figure 31

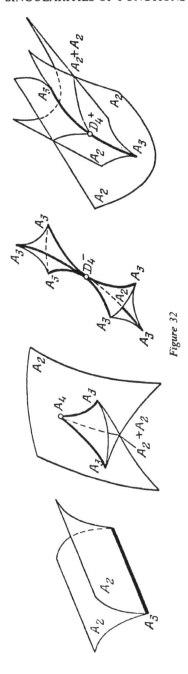

Figure 32

Parameters that go over into each other under a local diffeomorphism correspond to integrals of the same order.

For example, if the space is illuminated by a surface, the illumination of a point in general position is of order λ, where λ is the wavelength emitted by the surface. The illumination of the points of some surfaces in space (surfaces of caustics) have order $\lambda^{5/6}$. The illumination of particular curves (cuspidal edges of a caustic) have order $\lambda^{3/4}$. Individual points of the caustic—the vertices of swallowtails, pyramids and purses have illumination of order $\lambda^{7/10}$, $\lambda^{2/3}$, and $\lambda^{2/3}$, respectively.

2.1.6 Oscillatory Integrals over a Half-Space

2.1.6 Oscillatory Integrals over a Half-Space   We return to the example in par. 2.1.1. Assume that the radiating surface is opaque to the emitted waves. The waves from the visible parts of the surface arrive at a selected point of the space. Thus the total oscillation at the space point is expressed as the sum of oscillatory integrals, each of which is taken over part of the surface. Thus, in studying short-wave oscillations it is important to know how to calculate the asymptotics of oscillatory integrals over regions with boundaries. Let us examine the case of a smooth boundary.

Consider an oscillatory integral over part of the space $\mathbb{R}^n$ given by the condition that the first coordinate be positive, where the phase and amplitude are assumed to be smooth functions over all space.

In this situation we have the principle of stationary phase: if the phase of the oscillatory integral has no critical points in the integration half-space, and if the restriction of the phase to the boundary of the half-space also has no critical points, then as the parameter τ of the oscillatory integral goes to infinity, the integral tends to zero faster than any power of the parameter.

Assume that the phase of the oscillatory integral over the half-space has no critical points on the boundary of the half-space. Assume that all its critical points inside the half-space are nondegenerate and that the critical points of its restriction to the boundary of the half-space are also nondegenerate. We shall call such an oscillating integral a *Fresnel integral* over a half-space.

THEOREM 3 (see Ref. 10)   Assume that the phase of a Fresnel integral over a half-space has no critical points in the half-space of the integration (right up to the boundary). Assume also that the restriction of the phase to the boundary has a single, nondegenerate, critical point at the origin. Then as the parameter τ of the integral tends to $+\infty$, the integral is representable in the form

$$\varphi(0)(2\pi/\tau)^{(n-1)/2}(1/i\tau)\exp(i\tau f(0) + (i\pi/4)\mathrm{Sign}\,\tilde{f}''_{x'x'}(0))$$

$$\cdot|\det \tilde{f}''_{x'x'}(0)|^{-1/2} + O(\tau^{-(n+1)/2-1}),$$

where $\tilde{f}''_{xx}$ is the matrix of second partial derivatives of the restriction of the phase to the boundary.

Let us assume that the phase and amplitude of the oscillatory integral over the half-space depend on additional parameters. Assume that the phase, considered as a family of functions which depends on the parameters, is a family of functions in general position. In this case, the integral is a Fresnel integral for almost all values of the parameters. The set of those values for which the integral is not of Fresnel type forms a hypersurface in the parameter space, called a *caustic*.

Assume that for a selected value of the parameters the phase has a single critical point on the boundary of the half-space of integration. A caustic in the neighborhood of such a value of the parameters is called *elementary*.

Examples of elementary caustics occurring for the cases of 2 and 3 parameters are shown in Figs. 33 and 34. For a caustic value of the parameters either the phase has a degenerate critical point in the half-space of integration or the restriction of the phase to the boundary has a degenerate critical point. In the figures, near each part of

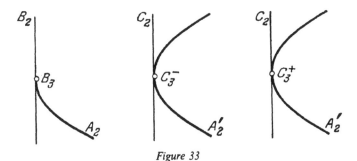

Figure 33

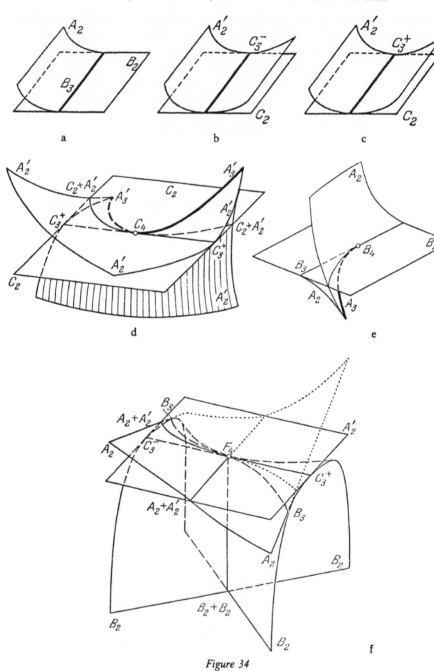

Figure 34

the caustic we give the notation for these degenerate points. For the normal forms of the critical points designated on the caustics, see Section 1, and also in Refs. 9 and 12. Each degenerate critical point contributes to the integral of order $\tau^{\beta - n/2}$. The number β for critical points of types A_μ', D_μ', B_μ, $C_\mu^\pm$ and F_4 are equal, respectively, to $-1(\mu + 1), 1/(\mu - 1), (\mu - 1)/2\mu, 0, \frac{1}{4}$ (see Ref. 13).

THEOREM 4 (see Refs. 9 and 12) For the phases of oscillatory integrals over a half-space, which depend in general fashion on two or three parameters, each elementary caustic is locally diffeomorphic to one of the caustics shown in Figs. 33 and 34. The parameters that go over into each other under a local diffeomorphism correspond to integrals of equal order of magnitude.

2.1.7 Zones of Light, Shadow, and Penumbra (according to P. K. Mandrykin) Let us assume that the phase and amplitude of the oscillatory integral depend on additional parameters. Consider the parameter space and a caustic located in it. Consider an arbitrary value of the parameters outside the caustic. The phase of the oscillatory integral corresponding to these parameters has only nondegenerate critical points, or has no critical points at all. In the first case, the oscillatory integral is of order $\tau^{-n/2}$, where n is the dimension of the space of integration; in the second case, as $\tau \to +\infty$ the oscillatory integral tends to zero faster than any power of the parameter τ. In this context, we call the region outside the caustic a light region if the values of the parameters for this region correspond to an oscillatory integral whose phase has at least one critical point; we call this region a region of shadow, if the parameters for this region correspond to oscillatory integrals whose phase has no critical points.

Example: In the pictures of caustics corresponding to critical points of types A_2, A_4, and D_4^+, the shadow regions are located below the caustics. The remaining regions outside the caustics, corresponding to critical points of types A_2, A_3, A_4, and $D_4^\pm$, are zones of light.

Let us now assume that the oscillatory integral is an integral over a half-space. Consider an arbitrary parameter value outside the caustic. There are three possibilities:

(1) The phase of the integral which corresponds to a selected parameter value has at least one critical point in the half-space of integration. In this case the integral has order $\tau^{-n/2}$.

(2)   The phase of the integral which corresponds to the particular parameter value has no critical points in the integration half-space, but the restriction of the phase to the boundary of the half-space has at least one critical point. In this case the integral has order $\tau^{-(n+1)/2}$.

(3)   The phase of the oscillatory integral which corresponds to the particular parameter values has no critical points in the integration half-space, and its restriction to the boundary of the half-space also has no critical points. In this case the integral tends to zero faster than any power of τ as the parameter $\tau \to +\infty$.

In accordance with these three possibilities, the regions into which the caustic divides the space of the additional parameters are called *zones of light, penumbra, and shadow*.

Example: The zones of penumbra and shadow in the figures of caustics corresponding to the critical points B_3, $C_3^{\pm}$, C_4, B_4, and F_4 are arranged in the following way. In Fig. 34a the penumbra is above the caustic. In Fig. 34b it is below it. In Fig. 34c the shadow zone is above the caustic; and the penumbra is between the two sheets of the caustic. In Fig. 34d the penumbra is on one side of the plane of the caustic. In Fig. 34e the penumbra is on the upper right hand side of the plane of the caustic. In Fig. 34f the shadow is above the whole caustic; the penumbra is behind the caustic below the plane of the caustic; another penumbra region is on the right hand side between the plane of the caustic and the surface with an edge.

2.2   Oscillatory Integrals Corresponding to Caustic Values of the parameters

2.2.1   THEOREM 5 (on asymptotic expansion, see Refs. 8, 13, 7 and 39)   Assume that the phase of the oscillatory integral in Eq. (1) is an analytic function in the neighborhood of its critical point x^o. Then the oscillatory integral is expanded in an asymptotic series

$$\exp(i\tau f(x^o)) \sum_{\alpha} \sum_{k=0}^{n-1} a_{k,\alpha}(\varphi) \tau^{\alpha}(\ln \tau)^k, \qquad \text{as} \quad \tau \to +\infty, \qquad (2)$$

if the amplitude is equal to zero outside a sufficiently small neighborhood of this point. Here the parameter α runs through a finite set of arithmetic progressions, depending only on the phase and consisting of negative rational numbers.

Example (see Ref. 10):   For a one-dimensional oscillatory integral with phase $x^{\mu+1}$, the asymptotic series has the form

$$\exp(i\tau f(x^o)) \sum_{l=1}^{\infty} a_l \tau^{-l/(\mu+1)}.$$

Example (cf. [10]): For a nondegenerate critical point of the phase depending on n variables, the asymptotic series has the form

$$\exp(i\tau f(x^o)) \sum_{l=0}^{\infty} a_l \tau^{-n/2-l}.$$

In the asymptotic series for an oscillatory integral the phase and the amplitude are not on an equal footing: the phase determines the power of the parameter, while the amplitude determines the coefficients for the powers of the parameter. The dependence on the phase is the more important factor. In the example of par. 2.1.1 for the radiation of waves by a surface, the phase corresponds to the geometry of the surface, while the amplitude corresponds to the intensity of the radiation.

2.2.2 Singularity and Oscillation Indices The fundamental characteristic of the asymptotic series in Eq. (2) is the power of the parameter in the maximal term of the series.

The *oscillation index* of a critical point of the phase is the maximum number α for which some of the coefficients $a_{k,\alpha}(\varphi)$ are not equal to zero for some amplitudes φ located in a sufficiently small neighborhood of the critical point of the phase.

Example: The oscillation index of a critical point of the phase $\pm x_1^{\mu+1} \pm x_2^2 \ldots \pm x_n^2$ is $-1(\mu+1) - (n-1)/2$.

The oscillation index of the critical point x^o of the phase f for the variables $x_1, \ldots, x_n$ is greater by $\frac{1}{2}$ than the oscillation index of the critical point $(x^o, 0)$ of the phase $f \pm z^2$ for $(n+1)$ variables. To eliminate the dependence on quadratic terms, we use the notion of a singularity index.

The *singularity index* of a critical point of the phase is the oscillation index plus $n/2$.

The singularity indices of the critical points of the phases f and $f \pm z^2$ are equal. The singularity index of a nondegenerate critical point is zero.

2.2.3 Tables of Singularity Indices (see Refs. 7 and 13) Below we give in tabular form the singularity indices of the critical points classified in Section 1. In the first row we give the designation of the critical point, in the second row we give the singularity index. The meaning of the tables is the following: if, in the neighborhood of a critical point, the phase is brought to the tabulated form by changes of the integration variables, then its singularity index is equal to the singularity index of the tabulated function.

Table III

A_k	D_k	E_6	E_7	E_8
$\dfrac{k-1}{2k+2}$	$\dfrac{k-2}{2k-2}$	$\dfrac{5}{12}$	$\dfrac{4}{9}$	$\dfrac{7}{15}$

Simple Singularities

$$P_8, X_9, J_{10}, J_{10+k}, X_{9+k}, Y_{r,s}, \tilde{Y}_r, P_{8+k}, P_{l,m}, \tilde{R}_m, T_{p,q,r}, \tilde{T}_{p,m}$$

$$\frac{1}{2}$$

E_{12}	E_{13}	E_{14}, Q_{10}	Z_{12}	Z_{13}, Q_{11}	W_{12}	W_{13}, S_{11}	Q_{12}	S_{12}	U_{12}
$\dfrac{11}{21}$	$\dfrac{8}{15}$	$\dfrac{13}{24}$	$\dfrac{6}{11}$	$\dfrac{5}{9}$	$\dfrac{11}{20}$	$\dfrac{9}{16}$	$\dfrac{17}{30}$	$\dfrac{15}{26}$	$\dfrac{7}{12}$

Unimodal Singularities

From these tables and the classification of critical points given in Section 1, it follows that (see Ref. 6):

For $l \leqslant 10$ and $k \geqslant 3$, the highest singularity index unremovable in families in general position of phases of k arguments which depend on l parameters has the form

$$\beta_l = \frac{1}{2} - \frac{1}{N}$$

where the number N is given by the Table:

l	0	1	2	3	4	5	6	7	8	9	10, $k=3$	11, $k=3$	10, $k>3$
N	$+2$	$+3$	$+4$	$+6$	$+8$	$+12$	∞	∞	-24	-16	-12	-8	-6

For example, for phases which depend in general manner on one, two, or three parameters, the order of the oscillatory integral for an arbitrary value of the parameters is no higher than $\tau^{\beta - n/2}$, where the numbers β are $1/6$, $1/4$ and $1/3$, respectively.

2.2.4 Formulation of Results and the Newton Polyhedra Let us formulate rules allowing us to calculate the oscillation indices. These rules are formulated in terms of the Newton polyhedra for the Taylor series of the critical points of the phase.

The Newton polyhedron is a convex polyhedron formed by the exponents of monomials that appear in the Taylor series.

If we consider a class of critical points of the phase with a fixed Newton polyhedron, it turns out that almost all critical points of the class have the same oscillation index. We give a formula expressing this common oscillation index in terms of the geometry of the Newton polyhedron. Exceptions are critical points at which the coefficients of the Taylor series satisfy explicitly written algebraic conditions.

It is useful to consider a class of critical points with a fixed Newton polyhedron in studying the discrete characteristics of critical points. As a rule, the characteristic takes a single value for almost all points of the class, this common value being expressed simply in terms of the geometry of the Newton polyhedron (see Section 3).

2.2.5 Definition of the Newton Polyhedron Consider the power series $f = \sum a_k x^k$ in n variables, where $k = (k_1, \ldots, k_n)$, and $x^k = x_1^{k_1} \ldots x_n^{k_n}$. We mark in the space $\mathbb{R}^n$ those points k with integer coordinates for which the coefficient a_k is not zero. All these points lie in the positive octant, i.e., the octant consisting of points with nonnegative coordinates. At each of these points we draw a parallel positive octant. The *Newton polyhedron of the series* is the convex hull of the union of all these octants.

Examples: The Newton polyhedra for the series

$$f = x_1^4 + 2x_1^2 x_2^2 + x_2^4 + x_2^5, \qquad g = x_1^4 - 2x_1^2 x_2^3 + x_2^6 + x_1^3 x_2^3,$$

$$h = x_1^4 + x_1^2 x_2^2 + x_2^6$$

are shown in Fig. 35.

C

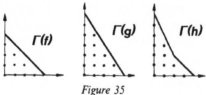

Figure 35

2.2.6   Distance to the Polyhedron and Remoteness of the Polyhedron
Consider the bisectrix of the positive octant in $\mathbb{R}^n$, i.e., the line
consisting of points with equal coordinates. The bisectrix (diagonal)
intersects the boundary of the Newton polyhedron at exactly one
point. This point is called the *center of the boundary* of the Newton
polyhedron. A coordinate of the center is called the *distance to the
Newton polyhedron*. The *remoteness of the Newton polyhedron* is the
reciprocal of the distance, taken with a minus sign.

Example: The distances to the Newton polyhedra for the series f,
g, and h of the preceding example are, respectively, 2, 12/5, and 2;
their remotenesses are, respectively, $-1/2$, $-5/12$, and $-1/2$.

The further the Newton polyhedron is located from the coordinate
origin, the greater its remoteness. We say that a Newton polyhedron is
remote if its remoteness is greater than -1. In other words, a Newton
polyhedron is remote if it does not include the point $(1, \ldots, 1)$.

2.2.7   Principal Part of a Power Series; Its Nondegeneracy   Let us
consider a power series f and its Newton polyhedron.

For any face γ of the Newton polyhedron we call γ-part of the
series that part consisting of monomials whose powers are on the face
γ; each monomial enters with the same coefficient that it had in the
original power series. If the face γ is bounded, then the γ-part is a
polynomial. We denote the γ-part of the series by f_γ.

The *principal part* of a power series is the polynomial consisting of
all monomials whose powers belong to bounded faces of the Newton
polyhedron; here the monomials enter with the same coefficients as
they had in the original power series.

Example: For the series f, g, and h, of the preceding example, the
principal parts are, respectively, the polynomials

$$x_1^4 + 2x_1^2 x_2^2 + x_4^2, x_1^2 - 2x_1^2 x_2^3 + x_2^6, x_1^4 + x_1^2 x_2^2 + x_2^6.$$

The principal part of a power series is said to be *nondegenerate* if for any closed bounded face γ of the Newton polyhedron of the series, the polynomials $\partial f_\gamma/\partial x_1, \ldots, \partial f_\gamma/\partial x_n$ have no common zeros in $(\mathbb{R}\backslash 0)^n$ (i.e., in the complement of the union of the coordinate hyperplanes).

Example: The principal parts of the series f, h of the preceding example are nondegenerate, while the principal part of the series g of the preceding example is degenerate.

It has been proved that there are few series with degenerate principal parts. Namely, *the set of degenerate principal parts is a proper algebraic subset in the space of all principal parts, corresponding to a given polyhedron.*

2.2.8 The Oscillation Index of a Critical Point of the Phase is Determined by the Remoteness of the Newton Polyhedron of its Taylor Series

THEOREM 6 (see Refs. 7 and 13) Suppose that the phase is an analytic function in the neighborhood of its critical point with zero critical value. Assume that the principal part of the Taylor series of the phase at this critical point is nondegenerate, and that the Newton polyhedron of the series is remote. Then the oscillation index of the critical point of the phase is equal to the remoteness of the Newton polyhedron.

Example: The degenerate critical point of the phase $x_1^{k_1} + x_2^{k_2}$ satisfies the conditions of the theorem. Its oscillation index is equal to $-1/k_1 - 1/k_2$.

Example: The critical points at the coordinate origin of the phases f and h in the example on p. 55 satisfy the conditions of the theorem. Their oscillation indices are, respectively, $-\frac{1}{2}$, $-\frac{1}{2}$.

According to Theorem 6, the oscillation index of a critical point of the phase can be expressed in terms of the Newton polyhedron of its Taylor series, if the principal part of its Taylor series is nondegenerate, and if the Newton polyhedron of the Taylor series is remote. A coordinate system in which the Taylor series has such properties does not always exist. For example, it does not exist for the critical point at the origin of the function g of the example on p. 55. Nevertheless, for the critical points of the phase depending on two variables one can drop the assumption of the existence of such a coordinate system.

By the *remoteness of a critical point* of the phase with zero critical

value we mean the maximum remoteness of the Newton polyhedra of
the Taylor series of the phase at the critical point in all systems of
local analytic coordinates in the neighborhood of the critical point.

A system of coordinates is said to be *adapted* to a *critical point* of
the phase if the remoteness of the Newton polyhedron of the Taylor
series of the phase at the critical point in this system of coordinates
has the maximum possible value, equal to the remoteness of the
critical point.

THEOREM 7 (see Refs. 7 and 13) Suppose that the phase is an
analytic function of two variables in the neighborhood of its critical
point with zero critical value. Then the oscillation index of the critical
point is equal to its remoteness.

The following assertion (see Ref. 7) is useful for finding adapted
coordinate systems: A coordinate system is adapted to a critical point
of the phase of two variables if at least one of the following conditions
is satisfied:

(1)   The center of the boundary of the Newton polyhedron of the
Taylor series at the critical point in this coordinate system is a vertex
of the polyhedron.

(2)   The center of the boundary of the Newton polyhedron lies on
an unbounded edge of the Newton polyhedron.

(3)   The center of the boundary of the Newton polyhedron lies on
a bounded edge of the Newton polyhedron and neither the tangent or
cotangent of the angle formed by the edge and the coordinate axis in
$\mathbb{R}^2$ is equal to an integer.

Example: Consider the functions g, h of the example on p. 55.
The coordinate system x_1, x_2 is adapted to the critical point at the
origin for these functions [as shown, respectively, by points (3) and (1)
of the preceding assertion]. According to Theorem 7, the oscillation
indices are $-5/12$ and $-1/2$, respectively.

In par. 2.2.3 we gave tables of singularity indices of the critical
points classified in Section 1. The singularity indices of the critical
points in the table were calculated from the oscillation indices. The
oscillation indices were calculated using Theorems 6 and 7. Theorems
6 and 7 are applicable to the critical points classified in Section 1,
since each of the critical points listed there either is a nondegenerate
principal part of the Taylor series, or is a function of two variables,
written in an adapted coordinate system.

2.2.9 Gelfand–Leray Form In studying oscillatory integrals, the following device, which reduces a multidimensional oscillatory integral to a one-dimensional integral is very useful. Let us consider the oscillatory integral

$$\int_{\mathbb{R}^n} e^{i\tau f(x)} \varphi(x)\, dx_1 \ldots dx_n.$$

We reduce the integral to a multiple integral, in which we first integrate over a hypersurface of constant phase, and then over the remaining variable, the value of the phase. To do this, we introduce new variables in the integral, one of which is the phase.

We shall make two comments. First, the phase can be regarded as a variable only away from its critical points. Therefore, we remove from the space the union of hypersurfaces of critical levels of the phase. The union of these hypersurfaces has zero measure and does not affect the integral. Second, for the integration over the level hypersurfaces we do not need to know each of the other new variables. It is sufficient to know on the level hypersurfaces a form of the $(n-1)$-dimensional density, which after multiplication by the differential of the phase becomes the form of the spatial volume. Such a density form is called a *Gelfand–Leray form* and is denoted by $dx_1 \wedge \cdots \wedge dx_n/df$.

Thus, the oscillatory integral is transformed to the form

$$\int_{-\infty}^{+\infty} e^{i\tau t} \left(\int_{f=t} \varphi\, dx_1 \wedge \cdots \wedge dx_n/df \right) dt.$$

In this representation the oscillatory integral is the Fourier transform of the function given by the inner integral. The function of one variable, defined by the inner integral, is called the *Gelfand–Leray function.*

The Gelfand–Leray function is smooth away from the critical values of the phase. In the neighborhood of a critical value of the phase the Gelfand–Leray function is expanded in an asymptotic series of the form

$$\sum_{\alpha} \sum_{k=0}^{n-1} a_{k,\alpha}(t-t_0)^{\alpha}(\ln(t-t_0))^k.$$

Knowing the asymptotic series of the Gelfand–Leray function, we can determine the asymptotic series of the oscillatory integral. Conversely, the asymptotics of the oscillatory integral give information about the asymptotics of the Gelfand–Leray function.

2.2.10 Volume of the Set of Smaller Values Let us assume that the phase has an isolated minimum and that the minimum value of the phase is equal to zero. We also assume that the amplitude in the neighborhood of the point of the minimum is identically equal to one. We denote by J the Gelfand–Leray function, and consider a new function

$$V(t) = \int_0^t J(s)\, ds.$$

It is obvious that, for negative values of the argument, this function is equal to zero, while for small positive values of the argument this function is equal to the volume of the set of points at which the phase takes on values less than the given one. Thus the asymptotics of the function giving the volume of the set of smaller values determines the asymptotics of the oscillatory integral in the case where the phase has an isolated minimum, and the amplitude is equal to a constant in the neighborhood of a point of minimum phase.

We give the order of magnitude of the rates at which the volume of the set of smaller values tends to zero for the simplest isolated minimum points.

Table IV, prepared by V. A. Vasiliev (see Ref. 4), gives the normal forms of germs of smooth functions in the neighborhood of points of minima (to within addition of constants and positive definite quadratic forms in additional variables). The reduction to normal form is accomplished by a smooth change of the independent variables. The column labelled l in the table indicates the number of parameters of the family beyond which the points of a minimum of the particular type become unremovable by small variations of the family. In general families of functions with $l < 16$ parameters, minimum points which are not equivalent to those enumerated in the table do not occur.

For small positive t the leading term in the asymptotics of the

Table IV

Designation	Normal Form	Restrictions	l
A_{2k-1}	x^{2k}	$k \geqslant 1$	$2k-2$
$X_{1,0} = T_{2,4,4}$	$x^4 + ax^2y^2 + y^4$	$a > -2$ $a \neq 2$	7
$X_{1,2r} = T_{2,4,4+2r}$	$x^4 + x^2y^2 + ay^{4+2r}$	$a > 0, r \geqslant 1$	$7 + 2r$
$Y^1_{2r,2q} = T_{2,4+2r,4+2q}$	$x^{4+2r} + ax^2y^2 + y^{4+2q}$	$a > 0, r, q \geqslant 1$	$7 + 2r + 2q$
$\tilde{Y}^1_{r,r} = \tilde{T}_{2,4+r,4+r}$	$(x^2 + y^2)^2 + ay^{4+r}$	$a \neq 0, r \geqslant 1$	$7 + 2r$
$W_{1,0}$	$x^4 + (a + by)x^2y^3 + y^6$	$a(-1)^q < 0$	12
$W^{\#}_{1,2q}$	$(x^2 + y^3) + (a + by)x^2y^{3+q}$	$q \geqslant 1$	$12 + 2q$

Here a and b are real parameters.

volume $V(t)$ of the set of smaller values has the form

$$\text{const}(\ln t)^k t^{-\beta + n/2},$$

where the number β is given in the table:

A_{2k-1}	$X_{1,0}, X_{1,2r}, Y'_{2r,2q}, \tilde{Y}^1_{r,r}$	$W_{1,0}, W^{\#}_{1,2q}$
$(k-1)/2k$	$1/2$	$7/12$

The number k is equal to 1 for critical points $X_{1,2r}$ and $Y'_{2r,2q}$; for the other critical points $k = 0$.

We formulate a general theorem about the calculation of the rate of fall to zero of the volume of the set of smaller values.

THEOREM 8 (see Refs. 7, 4 and 13) Assume that an analytic function has an isolated minimum, and that the minimum value is equal to zero. Then as $t \to +0$ the function V for the volume of the set of smaller values is expanded in an asymptotic series:

$$\sum_{\alpha} \sum_{k=0}^{n-1} a_{k,\alpha} t^{\alpha} (\ln t)^k.$$

Here the parameter α runs through a finite number of arithmetic progressions that consist of positive rational numbers. If, in addition, we know that the Taylor series of the function at the point of the minimum has a nondegenerate principal part, then the power α of the maximal term of the asymptotic series is equal to minus the remoteness of the Newton polyhedron of the Taylor series.

2.2.11 Area of a Level Surface   Consider a function having an isolated minimum. Suppose that the minimum value of the function is zero. In this case the level hypersurface for a small positive value of the function is a compact manifold. Let us calculate the $(n-1)$-dimensional volume of the compact level manifolds. We are interested in the asymptotic of this volume as the level tends to zero.

Example: For an isolated minimum point of the function

$$x_1^{a_1} + \cdots + x_n^{a_n}$$

where $a_1 \leqslant a_2 \leqslant \cdots \leqslant a_n$ are positive even integers, the leading term of the asymptotic of the $(n-1)$-dimensional volume of the manifold at level t, as $t \to +0$, has the form const. t^α, where $\alpha = 1/a_2 + \cdots + 1/a_n$.

In Ref. 13 a general formula is given for calculating asymptotics of the $(n-1)$-dimensional volume of level manifolds as the level tends toward the critical level.

2.2.12 Set of Points with a Small Gradient   Another characteristic of a critical point, like the one considered above, is the rate of tendency to zero of the volume of the set of points in which the *length* of the gradient is smaller than a given value, as this value tends to zero.

In the neighborhood of a selected critical point of the function, for each small positive t we consider the volume $V(t)$ of the set of those points of the neighborhood at which the square of the length of the gradient is less than t. We are interested in the asymptotic of the volume as $t \to +0$.

Example: For critical points of types A_k, D_4, and D_k ($k > 4$) and E_6, E_7 and E_8, the leading term of the asymptotic series of the function V, as $t \to +0$, has the form const. $t^{-\alpha+n/2}(\ln t)^k$, where (k, α) are, respectively,

$$((k-1)/2k, 0), (1/2, 0), (1/2, 1), (7/12, 0), (5/8, 0), (5/8, 0).$$

This proves Table II (p. 42).

To calculate the asymptotics of the volume of the set of points with a small gradient, we can use Theorem 8, which is used for the square of the length of the gradient.

2.3   Comparison of Oscillatory Integrals for Different Values of the Auxiliary Parameters

2.3.1   Uniform Estimates   In addition to the asymptotics of individual oscillatory integrals, it is frequently useful to know uniform estimates of oscillatory integrals that depend on auxiliary parameters.

Suppose that $f: \mathbb{R}^n \to \mathbb{R}$ is a smooth function. A *deformation* of it is any smooth function $F: \mathbb{R}^n \times \mathbb{R}^l \to \mathbb{R}$, equal to the function f for zero value of the second argument.

We say that at the critical point x^o of the phase f there is a uniform estimate with index α, if for any deformation F of the phase f there is a neighborhood in $\mathbb{R}^n \times \mathbb{R}^l$ of the point $x^o \times 0$, such that for any smooth function φ equal to zero outside this neighborhood and for any positive ϵ, there is a number $C(\epsilon, \varphi)$ for which

$$\left| \int_{\mathbb{R}^n} e^{i\tau F(x,\lambda)} \varphi(x,\lambda)\, dx_1 \ldots dx_n \right| < C(\epsilon, \varphi)\tau^{\alpha + \epsilon}$$

for all positive τ. The lower bound of such numbers α is called the *uniform oscillation index* of the phase at the critical point.

It is obvious that the uniform oscillation index is no smaller than the individual index.

We can assume that the individual oscillation index is equal to the uniform oscillation index, i.e., that the oscillatory integral admits an estimate from above, which is uniform in the additional parameters, by a quantity which is proportional to the magnitude of the integral for the initial value of the additional parameters.

For this assumption to be valid it is necessary that the individual oscillation index be upper semicontinuous for continuous changes of the critical point. Namely, it is necessary that the oscillation index for a complicated critical point be no less than that for a simpler critical point obtained by the decomposition of the more complicated one. Analysis of the table of singularity indices and the known adjacent critical points, as classified in Section 1, shows that such semicontinuity holds for the critical points classified in Section 1.

THEOREM 9 For simple (Refs. 5, 40), parabolic (Ref. 50) and hyperbolic critical points of the series $T_{p,q,r}$ (Refs. 2, 3) the uniform oscillation index is equal to the individual index.

It was reported in Ref. 1 that the uniform oscillation index is equal to the individual index for all critical points of the phase for the case of two variables.

COROLLARY OF THEOREM 9 For critical points that are unremovable, in families of phases in general position, which depend on no more than seven parameters, the uniform oscillation index is equal to the individual index.

According to Theorem 9, as we move along the caustic corresponding to one of the critical points listed in the theorem, the intensity of the short-wave oscillation at the limit point is no less than the intensity at a nearby prelimit point. Surprisingly, this phenomenon does not occur for all caustics. Namely, there are examples of highly degenerate critical points of the phase, for which the uniform oscillation index is greater than the individual oscillation index (Ref. 7). The codimensions of the critical points of the examples are at most 73, i.e., such points are removable by a small deformation from the families of functions with numbers of parameters smaller than 73.

According to the constructed examples, there exist critical points and their deformations that have the following property. The individual oscillation index of the critical point for a particular value of the deformation parameter is smaller than the individual oscillation index of the critical point for the general value of the deformation parameter, i.e., the magnitude of the oscillatory integral for the particular parameter value is essentially smaller than the magnitude of the oscillatory integral for the general value of the parameter.

It is interesting *whether this phenomenon can be observed physically* as a subset of the caustic that is darker than its surroundings. As already mentioned, such a phenomenon is not observed on caustics in general position in low-dimensional spaces (Theorem 9 and its corollary).

2.3.2 Estimates in the Mean Consider an oscillatory integral that depends on auxiliary parameters:

$$I(\tau,\lambda) = \int_{\mathbb{R}^n} e^{i\tau F(x,\lambda)} \varphi(x,\lambda)\, dx_1 \ldots dx_n .$$

If the value λ of the parameters does not belong to a caustic, the integral is of Fresnel type and has order $\tau^{-n/2}$. If the value of parameters values belongs to a caustic, the order of magnitude of the integral is increased.

As the following result shows, the order of magnitude of the average energy of short-wave oscillation near the caustic is the same as outside the caustic, i.e., the portion of points with large energy of oscillation is insignificant.

THEOREM 10 (see Ref. 40) Denote by $\sum$ the set of critical points of the phase, i.e.,

$$\sum = \{(x,\lambda) \mid \partial f/\partial x_j(x,\lambda) = 0, j = 1, \ldots, n\}.$$

Assume that $\sum$ is a submanifold, i.e., the differentials $d(\partial F/\partial x_j)$, $j = 1, \ldots, n$ are linearly independent at points of the set $\sum$. Assume that the amplitude of the oscillatory integral is equal to zero outside a sufficiently small neighborhood of one of the points of the set $\sum$. Then, as $\tau \to +\infty$ we have the asymptotic expansion

$$\int |I(\tau,\lambda)|^2 \, d\lambda \approx \sum_{l=n}^{\infty} a_l \tau^{-l},$$

where the numerical coefficients depend on the amplitude; in particular, the highest coefficient a_n is proportional to the integral of the square modulus of the amplitude over the critical set $\sum$.

Example: The versal deformation F of a finite multiplicity critical point satisfies the conditions of Theorem 10.

2.3.3 Number of Integer Points in a Region In recent years the use of the theory of singularities has been extended to include a number of new applications to problems of interaction of latticelike and smooth objects. One of the areas in which such interactions occur is the theory of resonances in nonlinear systems. A resonance between frequencies ω_k is an integer relation $\sum n_k \omega_k = n_0$. Such a relation determines a hypersurface in the space of frequencies. A system that depends on parameters is represented as a surface in the space of frequencies (its dimensionality, which depends on the number of parameters, is assumed to be smaller than the number of frequencies).

The question of how close a resonance of intermediate order is to a typical point of the hypersurface is important for the theory of nonlinear oscillations. The answer to it depends on the location of the surface relative to its tangent plane (inflexions and flatness of the surface cause sticking at resonances). For example, estimates of the rate of evolution of action variables of Hamiltonian systems that are close to integrable (N. N. Nekhoroshev) depend on the "steepness" of the unperturbed Hamiltonian function.

The area described—the theory of Diophantine approximations on submanifolds—has not yet been developed sufficiently, but it is already clear that the theory of singularities of wave fronts (i.e., singularities of Legendre transformations) plays a fundamental role in it.

Another area of interaction of smooth and integer-valued structures is the asymptotics of the number of points close to a smooth surface. Suppose that G is a region of volume V in Euclidean space $\mathbb{R}^n$. Then λG is the region G stretched out λ times, $N(\lambda)$ is the number of integer points in λG, $R(\lambda) = \lambda^n V - N(\lambda)$ is the error in the asymptotic formula $N(\lambda) \approx \lambda^n V$ ($\lambda \to \infty$). Estimating $R(\lambda)$ is a classical problem, which was studied by Gauss but which has not been solved even for a circle (where it is conjectured that R is of order $\lambda^{+1/2}$).

In this area the modern theory of oscillatory integrals leads to new results. In order to understand their significance, we note first that $|R(\lambda)| \leqslant c\lambda^{n-1}$ (this is how the area of the boundary of λG increases) and that the flatnesses of the boundary will increase R. For a sphere with center at 0, $|R(\lambda)| \geqslant c\lambda^{n-2}$ for certain sufficiently large λ (the number of integer points in the layer between spheres of radii λ and $\lambda + d$ is of order $\lambda^{n-1}d$, while the number of spheres containing integer points is of order λd, so that there is a sphere that contains no fewer than $c\lambda^{n-2}$ points).

Using the classification of singularities, Colin de Verdier proved the estimate $|R(\lambda)| \leqslant c\lambda^{n-2+2/n+1}$ for the region bounded by a smooth hypersurface in general position in $\mathbb{R}^n$ for $n \leqslant 7$ (Ref. 50).

The remainder R can be regarded as the sum of a large (order λ^{n-1}) number of terms with mathematical expectation zero (they correspond to cells cut out by the boundary of the region λG). If these terms were independent, their sum, from the laws of statistics, would be a quantity of order $\lambda^{(n-1)/2} \ll \lambda^{n-2}$ ($n > 3$).

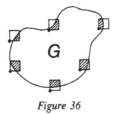

Figure 36

The theory of asymptotics of oscillatory integrals provides a basis for these arguments.

THEOREM 11 (see Refs. 30 and 31) The mean square value of $|R(\lambda)|$ over all lattices obtained from the lattice of integer points by rotations and displacements, admits an estimate from above of order $\lambda^{(n-1)/2}$ for any bounded region in $\mathbb{R}^n$ with a smooth boundary.

§3 The Geometry of Formulas

The number of integer points lying within a polyhedron, the volume of a polyhedron—these and other geometric quantities are encountered in answering various questions of the theory of singularities, in algebra and analysis. In this section we present examples of this kind.

3.1 The Newton Polyhedra

The Newton polyhedron of a polynomial which depends on several variables is the convex hull of the powers of the monomials appearing in the polynomial with nonzero coefficients. The Newton polyhedron generalizes the notion of degree and plays an analogous role. It is well known that the number of complex roots of a system of n equations of identical degree m in n unknowns is the same for nearly all values of the coefficients, and is equal to m^n (Bézout's theorem). Similarly, the number of complex roots of a system of n equations in n unknowns with the same Newton polyhedron is the same for nearly all values of the coefficients and is equal to the volume of the Newton polyhedron, multiplied by $n!$ (Kushnirenko's theorem, see 3.1.1).

The level line of a polynomial in two complex variables is a

Riemann surface. For nearly all polynomials of fixed degree n, the topology of this surface (the number of handles g) is expressed in terms of its degree, and does not depend on the values of the coefficients of the polynomial: $g = (n-1)(n-2)/2$. In the more general case, where instead of polynomials of fixed degree we consider polynomials with fixed Newton polyhedra, all the discrete characteristics of the manifold of zeros of the polynomial (or several polynomials) are expressed in terms of the geometry of the Newton polyhedra. Among these discrete characteristics are the number of solutions of a system of n equations in n unknowns, the Euler characteristic, the arithmetic and geometric genus of complete intersections, and the Hodge number of a mixed Hodge structure on the cohomologies of complete intersections.

The Newton polyhedron is defined not only for polynomials but also for germs of analytic functions. For germs of analytic functions in general position, with given Newton polyhedra, one can calculate the multiplicity of the zero solution of a system of analytic equations, the Milnor number and ζ function of the monodromy operator, the asymptotics of oscillatory integrals (see section 2), the Hodge number of the mixed Hodge structure in vanishing cohomologies, and in the two-dimensional and multidimensional quasihomogeneous cases, one can calculate the modality of the germ of the function.

In the answers one meets quantities characterizing both the sizes of the polyhedra (volume, number of integer points lying inside the polyhedron) as well as their combinatorics (the number of faces of various dimensions, numerical characteristics of their contacts).

In terms of the Newton polyhedron one can construct explicitly the compactification of complete intersections, and the resolution of singularities by means of a suitable toric manifold.

Thus, the Newton polyhedra connect algebraic geometry and the theory of singularities to the geometry of convex polyhedra. This connection is useful in both directions. On the one hand, explicit answers are given to problems of algebra and the theory of singularities in terms of the geometry of polyhedra. We note in this connection that even the volume of the convex hull of a system of points is a very complicated function of their coordinates. Therefore, the formulation of answers in numerical terms is so opaque that without knowing their geometric interpretation no progress is possible. On the other hand, algebraic theorems of general character (the Hodge theorem on the index of an algebraic surface, the Riemann-Roch theorem) give significant information about the geometry of polyhedra. In this way one obtains, for example, a simple proof of the Aleksandrov-Fenchel inequalities in the geometry of convex bodies.

The Newton polyhedra are also met in the theory of numbers (Ref. 17) in real analysis (Ref. 29), in the geometry of exponential sums (Refs. 32, 60), in the theory of differential equations (Refs. 18, 19, 36). In this paragraph we present formulations of some theorems about Newton polyhedra. More details can be found in Refs. 4, 7, 15, 12-13, 15-29, 32-34, 41, 58, 60-61.

3.1.1 The Number of Roots of a System of Equations with a Given Newton Polyhedron According to Bézout's theorem, the number of nonzero roots of the equation $f(z) = 0$ is equal to the difference between the highest and the lowest powers of the monomials appearing in the polynomial f. This difference is the volume of the Newton polyhedron (in the present case, the length of a segment). In the following two paragraphs we present the generalizations of this theorem to the case of arbitrary Newton polyhedra.

Let us start with definitions. A *monomial* in n complex variables is a product of the coordinates to integer (possibly negative) powers. Each monomial is associated with its *degree*, an integer vector, lying in n-dimensional real space, whose components are equal to the powers with which the coordinate functions enter in the monomial. A *Laurent polynomial* is a linear combination of monomials. The *support of the Laurent polynomial* is the set of powers of monomials entering in the Laurent polynomial with nonzero coefficients. (The Laurent polynomial is an ordinary polynomial if its support lies in the positive octant.) The *Newton polyhedron* of the Laurent polynomial is the convex hull of its support. It is much more convenient to consider the Laurent polynomials not in C^n but in the $(C \setminus 0)^n$-dimensional complex space, from which all the coordinate planes have been eliminated. With each face Γ of the Newton polyhedron of the Laurent polynomial f we associate a new Laurent polynomial, which is called the *restriction of the polynomial to the face*, denoted by f^Γ and defined as follows: only those monomials appear in f^Γ that have powers lying in the face Γ, with the same coefficients that they have in f.

Now let us consider a system of n Laurent equations $f_1 = \cdots = f_n = 0$ in $(C \backslash 0)^n$ with a common Newton polyhedron. The restricted system, $f_1^\Gamma = \cdots = f_n^\Gamma = 0$ corresponding to each face Γ of the polyhedron. The restricted system actually depends on a smaller number of variables, and in the case of general position is incompatible in $(C \backslash 0)^n$. We say that a system of n equations in n unknowns with a common Newton polyhedron is *regular* if all the restrictions of this system are incompatible in $(C \backslash 0)^n$. The following theorem of Kushnirenko holds: The number of solutions in $(C \backslash 0)^n$, counted with their multiplicities, of a regular system of n equations in n unknowns with a common Newton polyhedron is equal to the volume of the Newton polyhedron multiplied by n!

Example: The Newton polyhedron of the polynomial of degree m in n unknowns is the simplex $0 \leqslant x_1, \ldots, 0 \leqslant x_n, \sum x_i \leqslant m$ (we assume that the polynomial contains all monomials of degree $\leqslant m$). The volume of such a simplex is $m^n / n!$. The number of roots of the total system of n equations of degree m in n unknowns, according to Kushnirenko's theorem, is equal to m^n. This answer agrees with Bézout's theorem. If the polynomial does not contain all monomials of degree less than or equal to m, then the Newton polyhedron can be smaller than the simplex, so the number of solutions, calculated from Kushnirenko's theorem, can be smaller than the number m^n calculated from Bézout's theorem. Because of the absence of the monomials, certain infinitely distant points may be roots of the system of equations. Bézout's theorem, which calculates the number of roots of the system in projective space, takes into account these parasitic roots, while Kushnirenko's theorem does not.

Remark: The proof of Kushnirenko's theorem is found in Ref 41. Another proof can be extracted from the recent theorem of Atiyah on symplectic actions of tori (Ref. 62). We associate with each n-dimensional, integer-valued, convex polyhedron a symplectic "Veronese manifold" and a set of n commuting Hamiltonians on it, for each of which the motion is periodic with a period independent of the trajectory. The "moment mapping" corresponding to these Hamiltonians, according to Atiyah's theorem, maps the Veronese manifold on some convex polyhedron. It is not difficult to calculate that this polyhedron coincides with the original polyhedron. The calculation of the number of roots of a typical system of equations with a given

Newton polyhedron reduces to determining the volume of the Veronese manifold, which is easily done with the help of Atiyah's theorem: This volume is proportional to the volume of the polyhedron. The Veronese manifold is constructed from the convex integer polyhedron as follows. Consider the projective "space of monomials" CP^{N-1}, the number of whose homogeneous coordinates is equal to the number N of integer points lying inside and on the faces of Δ. By a Δ-*Veronese manifold* we mean the image of the mapping $(C \backslash 0)^n \to CP^{N-1}$, which associates the point z of $(C \backslash 0)^n$ with the point of projective space whose homogeneous coordinates coincide with the values at the point z of monomials whose degrees lie within Δ and on its faces. Since the Δ-Veronese manifold is algebraic, the numbers of points of its intersection with the planes of complementary dimension are the same for almost all planes. According to the Buffon-Crofton formula this number is proportional to the volume of the manifold. But a plane of codimension n in the space of monomials CP^{N-1} corresponds to a system of n equations with the Newton polyhedron Δ. The number of points of intersection of a plane with the Δ-Veronese manifold is equal to the number of solutions of the system. Therefore, the number of solutions of the system is proportional to the volume of the Veronese manifold. Atiyah's theorem is used to calculate this volume. A real n-dimensional torus acts in the space $(C \backslash 0)^n$: each of the n coordinates can be multiplied by a number with absolute value equal to one. This action of the torus on $(C \backslash 0)^n$ could be carried over to the Δ-Veronese manifold and is symplectic there. Thus, Atiyah's theorem is applicable; it follows from this theorem that the volume of the Δ-Veronese manifold is proportional to the volume of the original Newton polyhedron Δ.

3.1.2 How Does One Find the Number of Solutions of a System of n Equations in n Unknowns with Different Newton Polyhedra? Here is the answer to this question for a system in general position with fixed Newton polyhedra: the number of solutions not lying on the coordinate planes is equal to the mixed volume of the Newton polyhedra, multiplied by $n!$ Below we shall give the definition of mixed volume and describe explicitly the conditions for degeneracy.

The *Minkowski sum* of two subsets of a linear space is the set of sums of all pairs of vectors, in which one vector of the pair lies in one subset and the second vector in the other. The product of a subset and a number can be determined in a similar manner. The Minkowski sum of convex bodies (convex polyhedra, convex polyhedra with vertices at integer points) is a convex body (convex polyhedron, convex polyhedron with vertices at integer points). The following theorem holds:

MINKOWSKI THEOREM   The volume of a body which is a linear combination with positive coefficients of fixed convex bodies lying in R^n is a homogeneous polynomial of degree n in the coefficients of the linear combination.

Definition   The *mixed volume* $V(\Delta_1, \ldots, \Delta_n)$ *of the convex bodies* $\Delta_1, \ldots, \Delta_n$ *in* R^n *is the coefficient in the polynomial* $V(\lambda_1 \Delta_1 + \cdots + \lambda_n \Delta_n)$ *of* $\lambda_1 \times \cdots \times \lambda_n$, *divided by* $n!$ (here $V(\Delta)$ is the volume of the body Δ).

The mixed volume of n identical bodies is equal to the volume of any one of them. The mixed volume of n bodies is expressed in terms of the usual volumes of their sums in the same way as the product of n numbers is expressed in terms of the n-th powers of their sums. For example, for $n = 2$,

$$ab = \tfrac{1}{2}\left[(a + b)^2 - a^2 - b^2\right]$$

$$V(\Delta_1, \Delta_2) = \tfrac{1}{2}\left[V(\Delta_1 + \Delta_2) - V(\Delta_1) - V(\Delta_2)\right]$$

Similarly, for $n = 3$,

$$V(\Delta_1, \Delta_2, \Delta_3) = \frac{1}{3!}\left[V(\Delta_1 + \Delta_2 + \Delta_3) - \sum_{i<j} V(\Delta_i + \Delta_j) + \sum V(\Delta_i)\right]$$

Example: Suppose that Δ_1 is the rectangle $0 \leqslant x \leqslant a$, $0 \leqslant y \leqslant b$ and Δ_2 is the rectangle $0 \leqslant x \leqslant c$, $0 \leqslant y \leqslant d$. The Minkowski sum $\Delta_1 + \Delta_2$ is the rectangle $0 \leqslant x \leqslant a + c$, $0 \leqslant y \leqslant b + d$. The mixed volume $V(\Delta_1, \Delta_2)$ is equal to $ad + bc$. The number $ad + bc$ is the permanent of the matrix $\left[\begin{smallmatrix} a & c \\ b & d \end{smallmatrix}\right]$ (the definition of the permanent differs from that of the determinant only in that all the terms in the permanent have a plus sign). In the multidimensional case the mixed volume of n parallelepipeds with sides parallel to the coordinate axes is also equal to the permanent of the corresponding matrix.

Let us consider a system of n Laurent equations $f_1 = \cdots = f_n = 0$, with Newton polyhedra $\Delta_1, \ldots, \Delta_n$. Below we define the regularity condition for such systems.

BERNSHTEIN'S THEOREM The number of solutions in $(C\setminus 0)^n$ (which take account of the multiplicity) of a regular system of n equations in n unknowns, is equal to the mixed volume of the Newton polyhedra of the equations of the system, multiplied by n!

Example: The number of roots of a general system of polynomial equations, in which the i-th variable enters in the j-th equation with a power no higher than $a_{i,j}$, is equal to the permanent of the matrix $(a_{i,j})$, multiplied by n!

Kushnirenko's theorem coincides with Bernshtein's theorem for equations with identical Newton polyhedra. We now proceed to the definition of a *regular system of equations*. We first define the truncations of a system associated with a function ξ. We take an arbitrary linear function ξ on the space R^n in which the Newton polyhedra lie. We denote by f^ξ the restriction of the Laurent polynomial f to that face of its Newton polyhedron on which the linear function ξ takes its maximum value. We associate the restricted system $f_1^\xi = \cdots = f_n^\xi = 0$ with the system of equations $f_1 = \cdots = f_n = 0$ and the linear function ξ. For a nonzero function ξ the restricted system actually depends on a smaller number of variables. Therefore, in the case of general position such a system is inconsistent in $(C\setminus 0)^n$. A given system of equations has only a finite number of restricted systems (if the polyhedra of all the equations coincide, then the truncations correspond to the faces of the common polyhedron). A system of n equations in n unknowns is said to be regular if its restrictions for all nonzero functions ξ are inconsistent in $(C\setminus 0)^n$. It is just such a system to which Bernshtein's theorem is applicable.

We note that if for each nonzero linear function ξ, the maximum in one of the polyhedra is attained at a vertex then the regularity conditions are automatically satisfied. [In this case the truncated system contains an equation which is contained in a monomial set equal to zero; this equation has no solutions in $(C\setminus 0)^n$.] For example, in the case of $n = 2$, the regularity conditions are satisfied automatically, if the two Newton polyhedra on the real plane have no parallel sides.

3.1.3   The Newton Polyhedron of the Germ of an Analytic Function   In this paragraph we discuss the calculation of the multiplicity of a root of a system of equations, the Milnor number, and the modality of a function in two variables, in terms of Newton polyhedra.

We begin with definitions. To each power series in n variables we associate its support, the set of powers of monomials appearing in the series with nonzero coefficients. To each point of the support we associate an octant with vertex at that point, and consisting of all points for which each coordinate is not smaller than the corresponding coordinate of the vertex. The *Newton polyhedron* of the series is the convex hull of the union of all octants with vertices on the support of the series. *The Newton diagram of the series* is the union of compact faces of the Newton polyhedron. A Newton polyhedron of the series is said to be *suitable* if it intersects all the coordinate axes.

By the *volume $V(\Delta)$ of a suitable polyhedron* we mean the volume of the (nonconvex) region between the origin and the faces of the polyhedron in the positive octant R_+^n. Minkowski's theorem holds for the volumes of suitable polyhedra: the volume of a linear combination, with positive coefficients, of suitable polyhedra (just like the usual volume of polyhedra) is a polynomial in the coefficients of the linear combination. Therefore, we can define *a mixed volume for suitable polyhedra*, which is a verbatim repetition of the usual definition of mixed volume (see par. 3.1.2). For example, in the plane the mixed volume $V(\Delta_1, \Delta_2)$ of two suitable polyhedra Δ_1, Δ_2 is defined by the formula

$$2V(\Delta_1, \Delta_2) = V(\Delta_1 + \Delta_2) - V(\Delta_1) - V(\Delta_2)$$

The following local variant of Bernshtein's theorem (see par. 3.1.2) holds.

THEOREM   The multiplicity of the origin as of a regular solution of a system of n analytic equations of n unknowns with suitable Newton polyhedra is equal to the mixed volume of the suitable Newton polyhedra, multiplied by $n!$

Example: The multiplicity of the origin as of a regular solution of the system $x_1^a + x_2^b = x_1^c + x_2^d = 0$ with positive powers a, b, c, d, with $ad \neq bc$, is equal to $\min(ad, bc)$.

We give a precise definition of regularity of the origin. The linear function ξ is said to be negative if all its values in the positive octant are negative (i.e., $\xi = \sum a_k, x_k, \ a_k < 0$). The restriction f^ξ of the analytic function f to negative functions ξ is its truncation to that face of the Newton polyhedron at which the function ξ attains its maximum. A null solution of a system of n analytic equations $f_1 = \cdots = f_n = 0$ with suitable Newton polyhedra at the origin is said to be *regular* if the restricted of the system, $f_1^\xi = \cdots = f_n^\xi = 0$, is inconsistent in $(C\backslash 0)^n$ for all negative functions ξ.

We now give the formula for the Milnor number μ of the germ of an analytic function. First, we shall give several definitions and notations. We say that the germ of an analytic function f with suitable Newton polyhedron Δ is Δ-*nondegenerate* if, for any compact face Γ of the Newton polyhedron, the system $z_1(\partial f^\Gamma / \partial z_1) = \cdots = z_n(\partial f^\Gamma / \partial z_n) = 0$ is inconsistent in $(C\backslash 0)^n$ (here f^Γ is the truncation of f to the face Γ). We give some geometric notations: for a suitable Newton polyhedron $\Delta \subset R^n$, we denote by Δ^I its intersection with the coordinate plane R^I in R^n, by $d(I)$ the dimension of this coordinate plane and by $V(\Delta^I)$ the $d(I)$-dimensional volume of the (nonconvex) region in the positive octant R_+^I between the origin and the boundary of Δ^I.

THEOREM For a Δ-nondegenerate function f with a suitable Newton polyhedron Δ, the Milnor number is

$$\sum d(I)! \cdot (-1)^{n-d(I)} V(\Delta^I) + (-1)^n$$

where the summation is taken over all intersections Δ^I of the suitable Newton polyhedron with the coordinate planes.

Example: For almost all functions of two variables with a given suitable Newton polyhedron, $\mu = 2S - a - b + 1$, where S is the area under the polyhedron, a and b are coordinates of the points of the polyhedron on the axes (see the figure).

The *modality* $m(f)$ of a function f is the dimension of the space of orbits of the group of diffeomorphisms in the space of convergent power series which has no free and no linear terms in the neighborhood of the orbit of the point f.

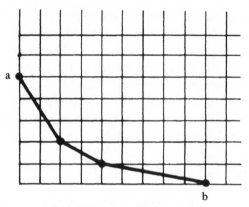

Calculation of the Milnor number.

$$\mu = 2S - a - b + 1$$

$$= 2 \cdot 12\tfrac{1}{2} - 5 - 9 + 1 = 12$$

THEOREM (KUSHNIRENKO) The modality of a Δ-nondegenerate function of two variables with suitable Newton polyhedron is equal to the number of integer points lying on the Newton diagram and below it, for which each coordinate is not less than two.

3.1.4 Complete Intersections Consider in $(C \backslash 0)^n$ a system of k Laurent equations $f_1 = \cdots = f_k = 0$ with Newton polyhedra Δ_1, $\ldots, \Delta_k$. For Laurent polynomials in general position with given Newton polyhedra, the discrete invariants of the set f solutions are identical and are expressed in terms of the polyhedra. Below we give an explicit description of the conditions for nondegeneracy and describe the Euler characteristic and the homotopy of the complete intersection.

Definition A system $f_1 = \cdots = f_k = 0$ is said to be *nondegenerate for its Newton polyhedra* if for any linear function ξ the following (ξ) condition is satisfied: for any solution z of the truncated system $f_1^\xi = \cdots = f_k^\xi = 0$, lying in $(C \backslash 0)^n$ the differentials df_i^ξ are linearly independent in the tangent space at the point z.

Each of the (ξ) conditions is satisfied for almost all Laurent polynomials with fixed polyhedra (this is the Sard-Bertini theorem).

SINGULARITIES OF FUNCTIONS 77

Although formally the nondegeneracy conditions are described for any linear function ξ, actually the different conditions are finite in number (there exist only a finite number of different truncated systems). Therefore the nondegeneracy conditions are satisfied almost everywhere. Among the nondegeneracy conditions there is the (0) condition (for the null function ξ), according to which the system $f_1 = \cdots = f_k = 0$ determines a nonsingular $(n - k)$-dimensional manifold in $(C\backslash 0)^n$. We note that for $k = n$ the class of regular systems is broader than the class of nondegenerate systems: for regular systems the (0) condition does not have to be satisfied (regular systems can have multiple roots).

THEOREM The Euler characteristic of the nondegenerate complete intersection $f_1 = \cdots = f_k = 0$ in $(C\backslash 0)^n$, $(k \leqslant n)$, with the Newton polyhedra $\Delta_1, \ldots, \Delta_k$ is equal to $(-1)^{n-k} \cdot n! \cdot \sum V(\Delta_1, \ldots, \Delta_k, \Delta_{i_1}, \ldots, \Delta_{i_{n-k}})$, where the sum is taken over all sets $1 = i_1 \leqslant \cdots \leqslant i_{n-k} \leqslant k$.

We note one special case of this theorem.

COROLLARY The Euler characteristic of a hypersurface in $(C\backslash 0)^n$ defined by a nondegenerate equation with fixed Newton polyhedron is equal to the volume of the Newton polyhedron multiplied by $(-1)^{n-1} \cdot n!$

Another consequence of this theorem is Bernshtein's theorem (see par. 3.1.2): for $k = n$, the nondegenerate complete intersections consist of points, and the Euler characteristic is equal to the number of points (strictly speaking, Bernshtein's theorem is slightly stronger than this corollary, since it is applicable to a degenerate regular system).

THEOREM (on homotopy type) Suppose that the Newton polyhedra Δ_i have a total dimension $\dim \Delta_i = n$ and the complete intersection $f_i = \cdots = f_k = 0$ in $(C\backslash 0)^n$ is nondegenerate. Then for $k < n$ the complete intersection is connected and has the homotopy type of a bouquet of $(n - k)$-dimensional spheres. The number of spheres in the bouquet is equal to $n! \sum V(\Delta_1, \ldots, \Delta_k, \Delta_{i_1}, \ldots, \Delta_{i_n}) - (-1)^{n-k}$ (the summation is taken over all sets $1 \leqslant i_1 \leqslant \cdots \leqslant i_{n-k} \leqslant k$).

COROLLARY The cohomology groups of a nondegenerate complete intersection with Newton polyhedra of complete dimension are different from zero only in dimensions $n - k$ and 0, where for $n > k$ the zero-measure cohomology group is one-dimensional.

3.1.5 Genus of Complete Intersections   The formulas given below for the genus of complete intersections are generalizations of the following formulas for Abelian and elliptic integrals. Let us consider a Riemann surface $y^2 = P_3(x)$ (the complex phase curve of motion of a point in a field with a cubic potential). This Riemann surface, which is diffeomorphic to the torus, exhibits a single, everywhere holomorphic, differential form dx/y (the differential of the time of motion along the phase curve). In the case of a potential of degree n, the curve $y^2 = P_n(x)$ is diffeomorphic to the sphere with g handles, where g is connected with n either by the formula $n = 2g + 1$, or the formula $n = 2g + 2$ (depending on the parity of n). The basis of holomorphic forms in this case is given by g forms of the type $x^m dx/y$, $0 \leqslant m < g$. The number g is the genus of the curve. The Newton polyhedron of the curve $y^2 = P_n(x)$ is a triangle with vertices $(0,0)$, $(0,2)$, and $(n,0)$. There are exactly g points with integer coordinates strictly inside this triangle. In terms of these points, one can give a basis for the space of holomorphic forms: the point $(1,a)$, which lies inside the triangle, corresponds to the form $x^{a-1} dx/y$. We give below the generalization of this procedure for constructing a basis of holomorphic forms for the multi-dimensional case.

The nondegenerate complete intersections $f_1 = \cdots = f_k = 0$, where the f_i are Laurent polynomials, are smooth algebraic affine manifolds. In the cohomologies of such manifolds there is an additional structure, namely, the mixed Hodge structure. The discrete invariants of such a structure are calculated in terms of the Newton polyhedra. We consider only the calculation of the arithmetic and geometric genus which are invariants of this kind.

We first recall some definitons and general statements. Suppose that Y is a nonsingular (possibly noncompact) algebraic manifold. The set of holomorphic p-forms on Y, which extend holomorphically to any nonsingular algebraic compactification, is automatically closed, and it realizes the zero class of homology of the manifold Y only if it is equal to zero. Forms of this kind form a subspace in the p-dimensional cohomologies of the manifold Y. We denote the dimension of this subspace by $h^{p,0}(Y)$. The *arithmetic genus* of the manifold Y is the alternating sum $\sum (-1)^p h^{p,0}(Y)$ of the numbers $h^{p,0}(Y)$. The *geometric genus* of the manifold Y is the number $h^{n,0}(Y)$, where n is the complex dimension of the manifold Y.

Now we turn to the Newton polyhedron. We shall use the *characteristic* $B(\Delta)$ *of integer polyhedra*. Here is its definition. Suppose that Δ is a q-dimensional polyhedron with vertices at integer points, lying in R^n and R^q is a q-dimensional subspace, containing Δ. The number $B(\Delta)$ is defined as the number of integer points lying strictly within the polyhedron Δ (in the geometry of the subspace R^q), multiplied by $(-1)^q$.

THEOREM   The arithmetic genus of the nondegenerate complete intersection $f_1 = \cdots = f_k = 0$ in $(C \setminus 0)^n$, $(k \leqslant n)$, with the Newton polyhedra $\Delta_1, \ldots, \Delta_k$ is

$$1 - \sum B(\Delta_i) + \sum_{j > i} B(\Delta_i + \Delta_j) - \cdots + (-1)^k B(\Delta_1 + \cdots + \Delta_k)$$

COROLLARY   The geometric genus of the nondegenerate complete intersection (for $k < n$) with the polyhedra of full dimensionality is

$$(-1)^{n-k}\left(-\sum B(\Delta_i) + \sum_{j > i} B(\Delta_i + \Delta_j) - \cdots \right.$$

$$\left. + (-1)^k B(\Delta_1 + \cdots + \Delta_k)\right)$$

PROOF OF THE COROLLARY   According to the theorem on the homotopic type of the complete intersection (see par. 3.1.4), $h^{0,0} = 1$ and $h^{p,0} = 0$ for $0 < p < n - k$. The corollary now follows from the calculation of the arithmetic genus.

We now give a complete discription of the holomorphic forms of highest dimension which could be extended holomorphically to the compactification, for the case of a nondegenerate hypersurface $f = 0$ in $(C \setminus 0)^n$ with the Newton polyhedron Δ of complete dimension. For each integer point, lying strictly within the Newton polyhedron Δ, we denote by w_a the n-form on the hypersurface $f = 0$, defined by the formula

$$w_a = z_1^{a_1} \times \cdots \times z_n^{a_n} \frac{dz_1}{z_1} \wedge \cdots \wedge \frac{dz_n}{z_n} \Big/ df; \qquad a = a_1, \ldots, a_n$$

THEOREM   The forms w_a lie in the space of forms extendable holomorphically on the compactification of the manifold $f = 0$; they are linearly independent and generate that space. In particular, the geometric genus of the hypersurface is equal to $|B(\Delta)|$.

Example: Consider a curve in the plane determined by the equation $y^2 = P_n(x)$. The interior monomials for the Newton polyhedron of this curve have the form yx^a, where $1 \leqslant a < n/2$. The forms w_a corresponding to the monomials coincide [on the curve $y^2 = P_n(x)$] with the forms $x^{a-1}dx/2y$. Thus, for the curve $y^2 = P_n(x)$ we get the usual description of all Abelian differentials.

3.1.6   Total Intersections in C^n   It is customary to consider systems of equations not in $(C \backslash 0)^n$, but in the usual complex space C^n. Calculations with Newton polyhedra also are carried out in this situation. The answers here are more messy. As an example we consider the calculation of the Euler characteristic of a hypersurface in C^n. Suppose that f is a polynomial in n complex variables with a nonzero free term and Newton polyhedron Δ. We introduce the following notation: Δ^I is the intersection of the Newton polyhedron Δ with the coordinate plane R^I in R^n, $d(I)$ is the dimension of this plane, and $V(\Delta^I)$ is the $d(I)$-dimensional volume of the polyhedron Δ^I.

THEOREM   Let f be a nondegenerate polynomial with the Newton polyhedron Δ which has a nonzero free term. Then $f = 0$ is a nonsingular hypersurface in C^n, intersecting transversally all the coordinate planes in C^n. The Euler characteristic of this hypersurface is equal to $\Sigma(-1)^{d(I)-1}d(I)! V(\Delta^I)$, where the summation runs over all intersections of the Newton polyhedron Δ^I with the (nonzero) coordinate planes. This theorem follows from the calculation of the Euler characteristic in $(C \backslash 0)^n$ (see par. 3.1.4) and from the additivity of the Euler characteristic. We also note that the formula given for a hypersurface is analogous to the formula for the Milnor number (see par. 3.1.3).

3.2   Index of a Vector Field

The investigation of the topology of real algebraic curves, surfaces, etc., is a necessary stage in the study of singularities in analysis (the Taylor expansion approximates all objects by algebraic ones, and the topology of these objects has a decisive influence over the topology of the analytic objects).

Unfortunately, very little is known about the topology of real algebraic manifolds (even curves) of high degree. Algebraic curves of degree 3 and 4 were studied by Newton and Descartes, but what a plane algebraic curve of degree 8 might look like is not known even today (a curve of sixth degree can't have more than 11 ovals, and in that case only one of them contains other ovals inside it, namely, 1, 5 or 9 others—this was discovered in 1970).

The theory of singularities enables one to give some restrictions on the topology of real algebraic manifolds. For example, from the expression for the Poincaré index of a singular point of the vector field in terms of the local algebra of the singular point there follows the inequality of Petrovskii-Oleinik for the Euler characteristic of algebraic manifolds of fixed degree.

3.2.1 *The Index of a Homogeneous Singular Point*   What values can the index of an isolated singular point of a vector field in R^n, whose components are homogeneous polynomials of degree $m_1, \ldots, m_n$, have? Theorem 1 formulated below gives a complete answer to this question.

We introduce the following notations. $m = m_1, \ldots, m_n$ is a set of natural numbers; $\Delta(m)$ is a parallelepiped in R^n, defined by the inequalities $0 \leqslant y_1 \leqslant m_1 - 1, \ldots, 0 \leqslant y_n \leqslant m_n - 1$; $\mu = m_1 \cdot \ldots \cdot m_n$ is the number of integer points in the parallelepiped $\Delta(m)$; $\prod(m)$ is the number of integer points in the central section $y_1 + \cdots + y_n = \frac{1}{2}(m_1 + \cdots + m_n - n)$ of the parallelepiped $\Delta(m)$.

THEOREM 1 (Refs. 33 and 34)   The index ind of an isolated singular point 0 of a vector field in R^n, whose components are homogeneous polynomials of degree $m_1, \ldots, m_n$, satisfies the inequality $|\text{ind}| \leqslant \prod(m)$ and the congruence ind $\equiv \mu \bmod 2$. There are no other restrictions on the index.

The proof of the estimate of the index in Theorem 1 is obtained from the signature formula for the index (see par. 3.2.5).

We note that the algebraic formula for the number $\prod(m)$ is by no means simple. Just the simple geometric interpretation of this number enables one to present examples of vector fields with all indices not forbidden by Theorem 1 (Ref. 33).

3.2.2   Petrovskii-Oleinik Inequalities   Consider a real nonsingular projective hypersurface A of degree $m + 1$, given by a homogeneous polynomial f in n variables.

The Petrovskii-Oleinik inequalities consist of the following:

$$|\chi(A) - 1| \leqslant N_n(m) \qquad \text{if } n \text{ is even;}$$

$$|\chi(B_+) - \chi(B_-)| \leqslant N_n(m) \qquad \text{if } n \text{ and } m \text{ are odd.}$$

Here χ is the Euler characteristic, B_+ and B_- are parts of RP^{n-1} given by the conditions $f \geqslant 0$, $f \leqslant 0$, respectively, while the number $N_n(m)$ is the number of integer points in the central section of the cube with side m [i.e., $N_n(m) = \prod(\underbrace{m, \ldots, m}_{n \text{ times}})$].

It is not difficult to show that both for even and odd n the left sides of the Petrovskii-Oleinik inequalities are equal in modulus to the index of the gradient of the homogeneous polynomial f giving the hypersurface. Thus, the Petrovskii-Oleinik inequalities represent a special case of the inequality $|\text{ind}| \leqslant \prod(m)$ of Theorem 1: this inequality must be applied to the index of the gradient vector field of the homogeneous polynomial of degree $m + 1$.

The Petrovskii-Oleinik inequalities were discovered in 1949 (Ref. 35). The relationship between these inequalities and the estimate of the index of a vector field was established in 1978 (Ref. 34).

3.2.3   Index of a Polynomial Field   What values can the total index of singular points of a vector field in the region R^n have, if the region is defined by a polynomial inequality $P_0 < 0$, and if we know the degrees of the components of the field and the degree of the polynomial P_0? In this paragraph we formulate a complete answer to this question.

We denote by ind the sum of the indices of all the singular points of the field V in R^n, and by ind$^+$ and ind$^-$ the sums of the indices of all the points of V in the regions $P_0 > 0$ and $P_0 < 0$. We shall say that the pair V, P_0 has a degree not exceeding (equal to) m, m_0, where $m = m_1, \ldots, m_n$, if the degree of the i-th component of the field does not exceed (is equal to) m_i, and the degree of the inequality $P_0 < 0$ does not exceed (is equal to) m_0. We shall say that the pair V, P_0 is *nondegenerate* if, first, the real hypersurface $P_0 = 0$ does not pass through singular points of the field V and if, secondly, the real singular points of the field V have finite multiplicity and lie "in a finite part of the space R^n" (the last condition means that the system $x_0 = \bar{P}_1 = \cdots = \bar{P}_n = 0$ has only the zero solution. Here the $\bar{P}_i$ are homogeneous polynomials of degree m_i in $n + 1$ variables, equal to P_i when $x_0 \equiv 1$.)

We supplement the geometric definitions of par. 3.2.1: $\prod(m, m_0)$ is the number of integer points of the parallelepiped $\Delta(m)$ satisfying the inequalities

$$\tfrac{1}{2}(m_1 + \cdots + m_n - n - m_0) \leqslant y_1 + \cdots + y_n$$

$$\leqslant \tfrac{1}{2}(m_1 + \cdots + m_n - n + m_0)$$

$O(m, m_0)$ is the number of integer points of the parallelepiped satisfying the inqualities

$$\tfrac{1}{2}(m_1 + \cdots + m_n - n - m_0) \leqslant y_1 + \cdots + y_n$$

$$\leqslant \tfrac{1}{2}(m_1 + \cdots + m_n - n)$$

We note that $O(m, m_0) = \tfrac{1}{2}(\prod(m, m_0) + \prod(m))$, and that $\prod(m) \equiv \prod(m, m_0) \equiv \mu \pmod{2}$.

THEOREM 2 (Ref. 33) For the nondegenerate pair V, P_0 with degrees m, m_0 the numbers $a = \text{ind}$, $b = \text{ind}^+ - \text{ind}^-$ and $c = \text{ind}^+$ satisfy the inequalities $|a| \leqslant \prod(m)$, $|b| \leqslant \prod(m, m_0)$, $|c| \leqslant O(m, m_0)$ and the congruences $a \equiv b \equiv c \equiv \mu \bmod 2$. Conversely, for any number a (number b, number c) satisfying these restrictions, there exists a nondegenerate pair V, P_0 of degrees m, m_0 for which $\text{ind} = a$ ($\text{ind}^+ - \text{ind}^- = b$, $\text{ind}^+ = c$).

We note that Theorem 1 is a special case of Theorem 2 (the case where the field V has homogeneous components and a single singular point at the origin).

We can also evaluate the index for vector fields V with "infinitely remote singular points." The number ind^+ is defined if the region $P_0 > 0$ contains only isolated singular points of the field V. The number ind is defined if all the singular points of the field V are isolated.

THEOREM 3 (Ref. 33)   Suppose that for the pair V, P_0 of degrees not exceeding m, m_0, the number ind^+ is defined. Then if $m_0 + \cdots + m_n \equiv n \pmod{2}$ then the modulus of the number ind^+ does not exceed $O(m, m_0)$. In this case no other restrictions exist on the number ind^+. If $m_0 + \cdots + m_n \not\equiv n \pmod{2}$ and m_0 is odd, then the modulus of the number ind^+ does not exceed $O(m, m_0 + 1)$. In this case there exists a pair V, P_0 with the extremal $\text{ind}^+ = \pm O(m, m_0 + 1)$.

COROLLARY   Suppose that V is a vector field of degree not exceeding $m = m_1, \ldots, m_n$ with isolated singular points. Then, for $m_1 + \cdots + m_n \equiv n \pmod{2}$ the estimate $|\text{ind}| \leqslant \prod(m)$ holds, while for $m_1 + \cdots + m_n \not\equiv n \pmod{2}$ we have the estimate $|\text{ind}| \leqslant O(m, 1)$. Both are exact.

3.2.4   Inverse Jacobian Theorem   To study a multiple singular point one uses a small deformation to dissociate it into nonmultiple singular points. The key information about such a dissociation is provided by the theorem on the inverse Jacobian, which is also of intrinsic interest. The inverse Jacobian theorem is applied, for example, in the proof of the signature formula for the index (see par. 3.2.5). We give a formulation of this theorem and derive the classic Euler-Jacobi classification formula from it.

Suppose that a system of n holomorphic equations in n complex unknowns depends on parameters, where for general values of the parameters it has only nonmultiple roots, while for exceptional values of the parameters (which form, in general, a hypersurface in parameter space) certain roots merge and become multiple. At a multiple root the Jacobian of the system is equal to zero, so when the nonmultiple roots of the system coalesce, the Jacobian of the system tends to zero at the coalescing roots. The reciprocal of the Jacobian is defined only at nonmultiple roots, and for coalescing roots tends to infinity. It turns out, however, that when one sums the reciprocals of the Jacobian over all the roots of the system, all the infinities cancel out, and the sum remains finite. We give an exact formulation of this theorem. Suppose that $U \subseteq C^n$ is a domain in C^n, $f: U \to C^n$ is a holomorphic mapping, $\mathscr{J} = \det(\partial f / \partial x)$ is the Jacobian of this mapping, V is a region in the image space that does not intersect the image of the boundary of the region U, and $h: U \to C$ is a holomorphic weight function.

THEOREM (about the inverted Jacobian)   The function

$$\varphi(a) = \sum_{f(x) = a} h(x) \mathscr{J}^{-1}(x) \qquad (1)$$

defined on regular values $a \in V$ of the mapping f, is extendable holomorphically to the whole region V [the summation in (1) is taken over all roots of the system $f(x) = a$ lying in the region U].

One of the corollaries of the inverted Jacobian theorem is the *Euler-Jacobi formula*. We recall this formula and give its derivation from the theorem.

Suppose that for the system

$$P_1 = \cdots = P_n = 0 \qquad (2)$$

of n polynomial equations of degrees $m_1, \ldots, m_n$ in n complex unknowns, all the roots are nonmultiple and "lie in the finite part" of the complex space. Suppose that h is an arbitrary polynomial in n variables, whose degree is less than the degree of the Jacobian $\mathscr{J}$ of the system, (i.e., less than $m_1 + \cdots + m_n - n$). In this case the Euler-Jacobi formula

$$\rho = \sum_{P(x) = 0} h(x) \mathscr{J}^{-1}(x) \qquad (3)$$

holds [where the summation goes over all roots of the system (2)].

The proof is obtained by applying Theorem 1 to a special mapping of the $(n + 1)$-dimensional space into itself with a special weight function. The first n components of this mapping are homogeneous polynomials in the $n + 1$ variables, which coincide with the polynomials of the system (2) on the hyperplane $x_{n+1} = 1$, the last component is the coordinate function x_{n+1}. The weight function is a homogeneous polynomial coinciding with the polynomial h on the hyperplane $x_{n+1} = 1$.

The inverse image of the point $(0, \epsilon)$ of this mapping consists of points $(\epsilon x, \epsilon)$, where x is a root of the system (2). A calculation shows that

$$\varphi(0,\epsilon) = \rho\epsilon^{p-(\sum m_i - n)}$$

where φ is the function defined by formula (1) for the mapping and for the weight function constructed above, ρ is a number defined by formula (3), and p is the degree of the weight function. As ϵ tends to zero, the function φ must remain bounded. If the degree of the weight polynomial is less than the degree of the Jacobian, then this is possible only if $\rho = 0$. Thus the formula of Euler-Jacobi is proved. We note that this formula is used in estimating the index of a polynomial vector field (see par. 3.2.3) and in the Petrovskii-Oleinik proof of their inequalities (Ref. 35).

3.2.5   General Theorems about the Index   In this paragraph we give a formulation of the signature formula for the index and prove the estimate of the index of a singular point of a field with homogeneous components from Theorem 1 of par. 3.2.1.

Suppose that $f : (R^n, 0) \to (R^n, 0)$ is a mapping with finite multiplicity and Q is the local algebra of this mapping. We denote by $\mathcal{J}$ the image of the Jacobian of the mapping f in the local algebra Q.

ASSERTION 1   The image of the Jacobian $\mathcal{J}$ in the local algebra is not equal to zero.

Suppose that l is an arbitrary linear real-valued function on the local algebra Q. We define a bilinear form B_l on the local algebra by the formula $B_l(a,b) = l(a \cdot b)$.

THEOREM 1 [signature formula for the index (Refs. 54 and 55)] For any linear function l that is positive on the image of the Jacobian $\mathcal{J}$, the signature of the bilinear form B_l is equal to the index of the singular point 0 of the mapping f.

COROLLARY   If in the μ-dimensional local algebra Q of the mapping f there exists an m-dimensional linear subspace L, the product of any two of whose elements is zero, then the modulus of the index of the point 0 does not exceed $\mu - 2m$.

PROOF OF THE COROLLARY   The space L is a zero space for the bilinear form B_l. The modulus of the signature of the quadratic form on a μ-dimensional space having a zero m-dimensional subspace does not exceed $\mu - 2m$.

Let us now prove the estimate of the index of a singular point with homogeneous components $m_1, \ldots, m_n$ given in par. 3.2.1.

As the generators of the local algebra of this singular point we may take monomials whose degrees in the i-th variable are strictly less than m_i for any $i \leq n$. Such monomials are in one-to-one correspondence with the integer points of the parallelepiped $\Delta(m)$ (see par. 3.2.1). Consider the R-linear subspace L, which is spanned by the monomials whose degrees are greater than half their possible maximum $[d > \frac{1}{2}(\sum m_i - n)]$. The product of any two monomials of this subspace is equal to zero in the local algebra.

Consequently, $|\text{ind}| \leq \mu - 2 \dim_R \overline{L}$.

The right side of this inequality is equal to the number of base monomials of average degree, i.e., is equal to $\prod(m)$ (see par. 3.2.1).

The estimate of Theorem 1 is proven. In conclusion, we give two general estimates of the index.

THEOREM 2   The index of a singular point of multiplicity μ in an n-dimensional space does not exceed $\mu^{1-1/n}$.

This theorem is derived from the signature formula using the Teissier inequalities (Ref. 59).

THEOREM 3 (Ref. 34)   The modulus of a singular point of a gradient vector field in an even number of variables does not exceed the number $h_1^{n/2,n/2}$.

The number $h_1^{n/2,n/2}$ in the theorem is one of the Hodge-Steenbrink numbers (Ref. 34), which characterizes the complex geometry of the germ of the hypersurface $f = 0$. For the gradient of a homogeneous function f the estimate of the theorem coincides with the estimate of Theorem 1 (see par. 3.2.1).

3.3 Few-Term Equations

The topology of objects given by algebraic equations (real algebraic curves, surfaces, singularities, etc.) becomes complicated rapidly as the degree of the equation increases. It has become clear recently that the complexity of the topology depends on the number of monomials

D

they contain, rather than on the degree of the equations: the theorems formulated below estimate the complexity of the topology of geometric objects in terms of how cumbersome their equations are.

3.3.1   Real Few-Term Equations   The following theorem of Descartes is well-known. The number of positive roots of a polynomial in one real variable does not exceed the number of changes of sign in the sequence of its coefficients (zero coefficients are dropped from the sequence).

COROLLARY (Descartes' estimate)   The number of positive roots of a polynomial is less than the number of terms in it.

A. G. Kushnirenko suggested that polynomials with a small number of terms be called fewnomials. Descartes' estimate shows that irrespective of the power ·of a few-term polynomial which may be arbitrarily large), it has few positive roots.

THEOREM 1 (Ref. 37)   The number of nondegnerate real roots of a system of n fewnomial equations in n unknowns, containing no more than k monomials (independent of the degree of the fewnomials) is estimated from above by a certain function φ_1 of n and k.

THEOREM 2 (Ref. 37)   The sum of the Betti numbers of the $(n - m)$-dimensional real manifold, determined by a nondegenerate system of m fewnomial equations, containing no more than k monomials, is estimated from above by some function φ_2 of n and k.

We give the familar estimate for the function $\varphi_1 : \varphi_1(n, k) < (n + 1)^k 2^{[k(k+1)/2]+n}$. This estimate is still far from exact. According to an unproved (but irrefutable) conjecture of A. G. Kushnirenko, an exact estimate of the number of solutions in the positive octant is

$$\prod_{1 \leqslant i \leqslant n} (k_i - 1),$$

where k_i is the number of monomials appearing in the i-th equation.

3.3.2   Complex Few-Term Polynomials   The complex roots of the simplest two-term equation $x^N - 1 = 0$, with increasing N, are uniformly distributed in argument. The theorem formulated below shows that the roots of any nondegenerate system of fewnomial equations are also uniformly distributed in argument.

Let us consider a regular system of equations $f_1 = \cdots = f_n = 0$ in $(C\backslash 0)^n$, with fixed Newton polyhedra $\Delta_1, \ldots, \Delta_n$ (see par. 3.1.2). In writing these equations we encounter only k monomials (i.e., the union of the supports of these equations contains only k points). We denote by $N(f, G)$ the number of solutions of this system, whose arguments lie in a given domain G of the torus $T = \{\varphi_1, \ldots, \varphi_n\}$ mod 2π. For the number $\prod(\Delta^*, \partial G)$ defined below, which depends only on the geometry of the Newton polyhedra and on the domain G, the following theorem is valid.

THEOREM (Ref. 60) There exists a function φ of n and k such that for every regular system of n equations with k monomials the relation

$$\left| N(f, G) - \frac{n!}{(2\pi)^n} V(G) V(\Delta_1, \ldots, \Delta_n) \right| \leqslant \varphi(n, k) \prod(\Delta^*, \partial G)$$

is valid. Here $V(G)$ is the volume of the region $G \subset T^n$, and $V(\Delta_1, \ldots, \Delta_n)$ is the mixed volume of the Newton polyhedra of the system.

We give the definition of the number $\prod(\Delta^*, \partial G)$. Suppose that Δ^* is the domain in R^n defined by the inequalities $\{\varphi \in \Delta^* \mid |m\varphi| < \pi/2$ for all integer-valued vectors m, lying in the union of the supports of the Laurent polynomials $f_1, \ldots, f_n\}$. The number $\prod(\Delta^*, \partial G)$ is defined as the minimum number of domains, which are equal to within a parallel displacement of the region Δ^*, necessary to cover the boundary ∂G of the region $G \subset T^n$.

Let us consider some special cases of the theorem: (1) the domain G coincides with the torus T^n. In this case $V(G) = (2\pi)^n$, $\prod(\Delta^*, \partial G) = 0$, and the theorem coincides exactly with Bernshtein's theorem (see par. 3.1.2); (2) the domain G contracts to a point $0 \in T^n$. In this case, $V(G) \to 0$, $\prod(\Delta^*, \partial G) = 1$, and the theorem coincides exactly with Theorem 1 (par. 3.3.1); (3) we simultaneously increase the polyhedra Δ_i (without overlapping them), without increasing the number of monomials k and without changing the domain G. In this case the mixed volume $V(\Delta_i, \ldots, \Delta_n)$ is of the order of the nth power of the size of the polyhedra Δ_i, while the number $\prod(\Delta^*, \partial G)$ is of the order of the $(n - 1)$-power of the size. In this case the theorem shows that the roots are distributed uniformly in argument.

3.3.3 Generalizations

The idea of few-term polynomials consists in the fact that the intersections of real level lines of "sufficiently simple" functions should be "sufficiently simple." At present, this notion has a basis not only for few-term polynomials, but also for real Liouville functions (Refs. 38 and 61), and for the even wider class of real, transcendent Pfaffian functions (Refs. 37 and 60). The class of Pfaffian functions is sufficiently wide: for example, if $f(x, y)$ is a Pfaffian function, then a solution of the differential equation $y' = f(x, y)$ is a Pfaffian function also. The construction of these functions is based on the fact that the solutions of the first-order equations do not oscillate (in contrast to the solutions of second-order equations such as $y'' = -y$).

The theorem on complex fewnomials has also been generalized (Ref. 60): This generalization allows one to use the theorem in estimating the number of zeros of the linear combinations of exponentials e^{ax}, where a is a real vector.

REFERENCES

1. Karpushkin, V. N., Uniform estimates of oscillating integrals in R^2, *Dokl. Akad. Nauk SSSR*, **254**, N1, 28–31 (1980).
2. Karpushkin, V. N., Uniform estimates of oscillating integrals, *Usp. Mat. Nauk* **36**, N4, 213 (1981).
3. Karpushkin, V. N., Uniform estimates of oscillating integrals with a parabolic or hyperbolic base. *Trud. sem. I. G. Petrovskii*, **9** (1983).
4. Vasil'ev, V. A., Asymptotic behavior of exponential integrals, Newton diagrams and classification of minimum points. *Funkts. Anal.* **11**, N3, 1–11 (1977).
5. Vinogradov, I. M., The method of trigonometric sums in the theory of numbers, Moscow, Nauka, 1971.
6. Arnol'd, V. I., Remarks on the method of stationary phase and Coxeter numbers, *Usp. Mat. Nauk* **28**, N5, 17–44 (1973).
7. Varchenko, A. N., Newton polyhedra and estimates of oscillating integrals. *Funkts. Anal.* **10**, N3, 13–38 (1976).
8. Bernshtein, I. N. and Gelfand, S. I., Meromorphy of the functions P^λ. *Funkts. Anal.* **3**, N1, 84–86 (1969).
9. Arnol'd, V. I., Critical points of functions on manifolds with boundary, simple Lie groups B_k, C_k, F_k, and singularities of evolutes. *Usp. Mat. Nauk* **13**, N5, 91–105 (1978).
10. Fedoryuk, M. V., Saddle point method, Moscow, Nauka 1977.
11. Landau and Lifshitz, Mechanics, Moscow, Nauka, 1965.

12. Arnol'd, Varchenko and Gusein-Zade, Singularities of differentiable mappings, Vol. 1, Moscow, Nauka 1982.
13. Arnol'd, Varchenko and Gusein-Zade, idem, vol. 2, Moscow, Nauka 1983.
14. Arnol'd, V. I., Catastrophe theory, Moscow, Znanie, 1981.
15. Bernshtein, Kushnirenko and Khovanskii, Newton polyhedra, *Usp. Mat. Nauk* **31** N3, 201–202 (1976).
16. Kushnirenko, A. G., The Newton polyhedron and the number of solutions of equations in n unknowns, *Usp. Mat. Nauk* **30**, 32, 302–303 (1975).
17. Chebotarev, N. G., The "Newton polyhedron" and its role in the contemporary development of mathematics. Collected works, vol. 3, Moscow-Leningrad, Acad. Press, 1950.
18. Brüno, A. D., Local method of nonlinear analysis of differential equations, Moscow, Nauka 1979.
19. Berezovskaya, F. S., Index of a stationary point of a vector field in the plane, *Funkts. Anal.* **13**, N2, 77 (1979).
20. Kushnirenko, A. G., Newton polyhedra and Milnor numbers, *Funkts. Anal.* **9**, N9, 74–75 (1976).
21. Bernshtein, D. N., Number of roots of a system of equations, *Funkts. Anal.* **9**, N3, 1–4 (1975).
22. Kushnirenko, A. G., Newton polyhedra and Bézout's theorem, *Funkts. Anal.* **10**, N3, 82–83 (1976).
23. Danilov, V. I., Netwon polyhedra and vanishing cohomologies, *Funkts. Anal.* **13**, N2, 32–47 (1979).
24. Khovanskii, A. G., Newton polyhedra and toric manifolds, *Funkts. Anal.* **11**, N4, 56–67 (1977).
25. Khovanskii, A. G., Newton polyhedra and genus of complete intersections, *Funkts. Anal.* **12**, N1, 51–61 (1978).
26. Khovanskii, A. G., Newton polyhedra and the Euler–Jacobi formula, *Usp. Mat. Nauk* **33**, N6, 245–246 (1978).
27. Danilov, V. I., Geometry of toric manifolds, *Usp. Mat. Nauk* **33**, N2, 85–135 (1978).
28. Khovanskii, A. G., Geometry of convex manifolds and algebraic geometry, *Usp. Mat. Nauk* **34**, N4, 160–161 (1978).
29. Khovanskii, A. G., Newton polyhedra and the index of a vector field, *Usp. Mat. Nauk* **36**, N4, 234 (1981).
30. Varchenko, A. N., On the number of integer points in a region, *Usp. Mat. Nauk* **37** (1982).
31. Varchenko, A. N., On the number of integer points in families of homothetic regions, *Funkts. Anal.* **17**, (1983).
32. Kazarnovskii, B. Ya., On the zeros of exponential sums, *Dokl. Akad. Nauk SSSR* **257**, N4, 804–808 (1981).
33. Khovanskii, A. G., Index of a polynomial vector field, *Funkts. Anal.* **13**, N1, 49–58 (1979).
34. Arnol'd, V. I., Index of singular points of a vector field, the Petrovskii–Oleinik inequalities and mixed Hodge structures, *Funkts. Anal.* **12**, N1, 1–14 (1978).
35. Petrovskii and Oleinik, On the topology of real algebraic surfaces, *Izv. Akad. Nauk SSSR. ser. mat.* **13**, 389–402 (1949).
36. Bogdanov, R. I., Local orbital normal forms of vector fields in the plane, *Trud. sem. I. G. Petrovskii*, N5, 51–84 (1975).
37. Khovanskii, A. G., A class of systems of transcendental equations, *Dokl. Akad. Nauk SSSR* **255**, N4, 804–807 (1980).

38. Gelfond, O. A., Khovanskii, A. G., On real Liouville functions, *Funkts. Anal.* **14**, N2, 52–53 (1980).
39. Atiyah, M. F., Resolution of singularities and division of distributions, *Comm. Pure Appl. Math.* **23**, N2, 145–150 (1970).
40. Duistermaat, J., Oscillatory integrals, Lagrange immersions and unfoldings of singularities, *Comm. Pure Appl. Math.* **27**, N2, 201–281 (1974).
41. Kouchnirenko, A. G., Polyèdres de Newton et nombres de Milnor, *Invent. Math.* **32**, 1–31 (1976).
42. Nye and Potter, The use of catastrophe theory to analyse the stability and toppling of icebergs, *Annals of Glagiology* **1**, 49–54 (1980).
43. Nye, Cooley and Thorndike, The structure and evolution of flow fields and other vector fields, *J. Phys. A: Math. Gen.* **11**, N8, 1455–1490 (1978).
44. Nye, J. F., The motion and structure of dislocations in wavefronts, *Proc. Roy. Soc. (London)* **A378**, 219–239 (1981).
45. Nye, J. F., Optical caustics from liquid drops under gravity: observations of the parabolic and symbolic umbilics, *Phil. Trans. Roy. Soc.* (London) **292**, 25–44 (1979).
46. Nye, J. F., Optical caustics in the near field from liquid drops, *Proc. Roy. Soc.* (London) **A361**, 21–41 (1978).
47. Berry and Upstill, Catastrophe optics: morphologies of caustics and their diffraction patterns, Progress in Optics, XVIII, North-Holland, 1980.
48. Berry, M. V., Singularities in waves and rays, Les Houches Summer School, 1980; Amsterdam, North-Holland, 1981.
49. Berry, M. V., Waves and Thom's theorem, *Adv. Phys.* **25**, 1–26 (1976).
50. Colin de Verdier, Y., Nombre de points entiers dans une famille homothétique de domains de R^n, *Ann. Scient. Ecole Norm. Super. ser. a* **10**, 559–575 (1977).
51. Nye and Thorndike, Events in evolving three-dimensional vector fields, *J. Phys. A. Math. Gen.* **13**, 1–14 (1980).
52. Randol, B., On the Fourier transform of the indicator function of a planar set, *Trans. AMS* **139**, 271–278 (1969).
53. Randol, B., On the asymptotic behaviour of the Fourier transform of the indicator function of a convex set, *Trans. AMS* **139**, 279–285 (1969).
54. Eisenbud and Levine, The topological degree of a finite C^∞-map germ, *Ann. Math.* **106**, N1, 19–38 (1977).
55. Khimshiashvili, G. N., On the local degree of a smooth mapping, *Repts. Acad. Georgian SSSR*, **85**, N2, 309–311 (1977).
56. Arnol'd, V. I., Mathematical Methods of Classical Mechanics, Springer Verlag, New York–Heidelberg–Berlin (1980).
57. Arnol'd, Shandarin and Zeldovich, The large scale structure of the universe, I. General properites; one- and two-dimensional models, Geophysical and Astrophysical Hydrodynamics (1982).
58. Teissier, B., Variétés toriques et polytopes, *Sem. Bourbaki*, 33e annee, N565, 1980/81.
59. Teissier, B., Appendix to Levine's–Eisenbud's article, *Ann. Math.* **106**, 39 (1977).
60. Hovansky, A., Sur les racines complexes de systèmes d'équations algébraiques ayant un petit nombre de monômes, *C. R. Acad. Sc. Paris* **292**, 937–940 (1981).
61. Khovanskii, A., Théoréma de Bézout pour les fonctions de Liouville, preprint IHES/M/81/45, Septembre, 1981, Bures-sur-Yvette, France.
62. Atiyah, M. F., Convexity and commuting hamiltonians, *Bull. London Math. Soc.* **14**, part 1, N46 (1982).

REVIEWS IN MATHEMATICS AND MATHEMATICAL PHYSICS

Editor:
A.T.Fomenko
Moscow State University
Russia

Aims and Scope
Reviews In Mathematics and Mathematical Physics publishes review papers covering significant developments in mathematics and mathematical physics from Russia and the former Soviet Union. Topics includee soliton theory, the theory of quantum topological models and their applications for algebraic and differential geometry and topology.

World Wide Web Addresses
Additional information is also available through the Publisher's web home page site at
http://www.cambridgescientificpublishers.com

Ordering Information
Each volume consists of an irregular number of parts depending upon extent. Issues are available individually as well as by subscription. 2012 Volume 14/15.
Orders may be placed with your usual supplier or at the address shown below. Journal subscriptions are sold on a per volume basis only. Claims for nonreceipt of issues will be honored if made within three months of publication of the issue. All issues are dispatched by airmail throughout the world.

Subscription Rates
Base list subscription price per volume: EUR 120.00.* This price is available only to individuals whose library subscribes to the journal OR who warrant that the journal is for their own use and provide a home address for mailing. Orders must be sent directly to the Publisher and payment must be made by check or credit card.
Separate rates apply to academic and corporate/government institutions.
*EUR (Euro). The Euro is the worldwide base list currency rate. All other currency payments should be made using the current conversion rate set by Publisher. Subscribers should contact their agents of the Publisher. All prices are subject to change without notice.

Orders should be placed through the Publisher at the following address:
Cambridge Scientific Publishers Ltd
45 Margett Street
Cottenham
Cambridge
CB24 8QY
UK
Tel: +44 (0)1954 251283
Fax: +44 (0)1954 252517
Email: janie.wardle@cambridgescientificpublishers.com
Website: www.cambridgescientificpublishers.com

Reviews in Mathematics and Mathematical Physics
Notes for Contributors

Manuscripts
Papers should be typed with double spacing and wide margins (3 cm) on good quality paper and submitted in triplicate. Authors may also submit papers on disk in any format. Papers should be sent to Professor A.T.Fomenko, Department of Mathematics and Mechanics, Moscow State University, Moscow 119899, Russia. Email: atfomenko@mail.ru
or to Janie Wardle, Cambridge Scientific Publishers, 45, Margett Street, Cottenham, Cambridge, CB24 8QY, UK. Email: janiewardle@cambridgescientificpublishers.com.

Submission of a paper to *Reviews in Mathematics and Mathematical Physics* will be taken to imply that it represents original work not previously published, that is not being considered for publication elsewhere, and that if accepted for publication it will not be published in the same language without the consent of the Editors and Publisher.

Language: The language of publication is English.

Abstract: Each paper requires an abstract of 100-150 words summarizing the significant coverage and findings. It is a condition of acceptance by the Editor of a typescript for publication that the Publishers acquire automatically the copyright in the typescript throughout the world.

Figures
All figures should be numbered with consecutive arabic numbers, have descriptive captions and be mentioned in the text. Keep figures separate from the text, but indicate an approximate position for each in the margin.

Preparation: Figures submitted must be of a high enough standard for direct reproduction. Line drawings should include all the lettering and symbols included. Alternatively, good sharp photoprints ("glossies") are acceptable. Photographs intended for half-tone reproduction must be glossy original prints of maximum contrast. Clearly label each figure with author's name and figure number, indicate "top" where this is not obvious. Redrawing or retouching of unusable figures will be charged to the authors.

Size: Figures should be planned so that they reduce to 11 cm column width. The preferred width of line drawings is 15 to 22 cm with capital lettering 4 mm high, for reduction by onehalf. Photographs for half-tone reproduction should be about twice the desired size.

Color plates: Whenever the use of color is an integral part of the research, or where the work is generated in color, the Journal will publish the color illustrations without charge to the author. Reprints in color will carry a surcharge. Please write to the Publisher for details.

Equations and Formulae
Any mathematical or chemical notation should be clearly marked.

Mathematical equations: Mathematical equations should preferably be typewritten, with subscripts and superscripts clearly shown. It is helpful to identify unusual or ambiguous symbols in the margin when they first occur. To simplify typesetting, please use: 1) the "exp" form of complex exponential functions; 2) fractional exponents instead of root signs; and 3) the solidus (/) to simplify fractions – e.g. $\exp x_{1/2}$.

Chemical: Ring formulae and other complex chemical matter are extremely difficult to typeset. Please, therefore, supply reproducible artwork for equations containing such chemistry.

Marking: Where chemistry is straightforward and can be set (e.g. single-line formulae), please help the typesetter by distinguishing between, e.g., double bonds and equal signs, and single bonds and hyphens, where there is ambiguity. The printer finds it extremely difficult to identify which symbols should be set in roman (upright) or italic or bold type, especially where the paper contains both mathematics and chemistry. Therefore, please underline all mathematical symbols to be set in italic and put a wavy line under bold symbols. Other letters not marked will be set in roman type.

Tables

Number tables consecutively with arabic numerals and give a clear descriptive caption at the top. Avoid the use of vertical rules in the tables. Indicate in the margin where the printer should place the tables.

References and Notes

References and Notes are indicated in the text by consecutive superior arabic numbers (without parentheses). The full list should be collected and typed at the end of the paper in numerical order. Listed references should be complete in all details but excluding article titles in journals. Authors' initials should follow their names; journal title abbreviations should conform to *Physical Abstracts* style.

Examples:

Smith, A.B. and Jones, C.D., 1990, *J. Appl. Phys.* 34, 296.

Brown, R.B. *Molecular Spectroscopy*, Gordon and Breach, New York, 1965, 3rd ed., Chap. 6, pp. 95-106.

Proofs

Contributors from the former Soviet Union will receive page proofs (including figures) for correction via our internal courier network to Moscow. These must be returned to our Moscow office (Victor Selivanov, Lebedev Physical Institute, 53 Leninsky Prospect, Moscow 117924, Russia. Email: victor.d.selivanov@gmail.com) within 48 hours of receipt. All other contributors will receive page proofs (including figures) by airmail for correction, which must be returned within 48 hours of receipt. Please ensure that an email address/ full postal address if given on the first page of the typescript, so that proofs are not delayed in the post. Author's alterations in excess of 10% of the original composition cost will be charged to authors.

Reprints

Additional reprints may be ordered by completing the appropriate form sent with proofs.

Page Charges

There are no page charges to individuals or institutions.

Lightning Source UK Ltd.
Milton Keynes UK
UKHW011837081022
410109UK00002B/366